Palgrave Studies in Translating and Interpreting

Series Editor
Margaret Rogers, School of Literature and Languages, University of Surrey, Guildford, UK

This series examines the crucial role which translation and interpreting in their myriad forms play at all levels of communication in today's world, from the local to the global. Whilst this role is being increasingly recognised in some quarters (for example, through European Union legislation), in others it remains controversial for economic, political and social reasons. The rapidly changing landscape of translation and interpreting practice is accompanied by equally challenging developments in their academic study, often in an interdisciplinary framework and increasingly reflecting commonalities between what were once considered to be separate disciplines. The books in this series address specific issues in both translation and interpreting with the aim not only of charting but also of shaping the discipline with respect to contemporary practice and research.

Sofia Malamatidou

Translating Tourism

Cross-Linguistic Differences of Alternative Worldviews

palgrave
macmillan

Sofia Malamatidou
Department of Modern Languages
University of Birmingham
Birmingham, UK

ISSN 2947-5740 ISSN 2947-5759 (electronic)
Palgrave Studies in Translating and Interpreting
ISBN 978-3-031-49348-5 ISBN 978-3-031-49349-2 (eBook)
https://doi.org/10.1007/978-3-031-49349-2

Cover illustration: Maciej Bledowski/Alamy Stock Photo

This Palgrave Macmillan imprint is published by the registered company Springer Nature Switzerland AG
The registered company address is: Gewerbestrasse 11, 6330 Cham, Switzerland

To Andreas & Alexandra

Acknowledgements

This book is the culmination of many years of work, and it would not have been possible without the help of many people who supported me along the way. The first people I am grateful to are the postgraduate students, Vivian Nioti who helped me collect the Greek materials, and Daisy Savage who worked on the French. Thanks also go to all my former students at the University of Birmingham who had to endure my early attempts at teaching tourism translation and understanding its challenges and also to colleagues from the universities of Athens and Thessaloniki, most notably Maria Sidiropoulou and Elpida Loupaki, who invited me to give talks on the subject, and the valuable interactions I had with their students. Also, I am grateful to Stavroula Kefala, the Director for Research at the Ministry of Tourism in Greece, and Elena Manca from the University of Salento, for their valuable insights and conversations, as well as their support and enthusiasm around potential projects exploring tourism translation further. They have given me confidence in the potential of this book when I most needed it. Elena's research in particular largely inspired how I approached tourism translation in the book, and I hope to continue to learn from her in the future. A really big thank you goes to my mentor and very good colleague, Jeannette Littlemore, for reading and commenting on various drafts of the book, for always supporting me, even when I doubted myself, and for her boundless enthusiasm about my ideas. Her untiring guidance is evident in almost every aspect of this book, and it really would have been a very different book without her input. I would also like to express my sincere appreciation to Margaret Rogers, the series editor, for her invaluable contribution to this book. Her expertise, keen insights, and meticulous attention to detail have greatly enhanced the quality of the final product. I am truly grateful for her dedication and support throughout the editing process. Finally, thanks are due to my family, who supported me in many different ways and gave me space and time to work on the book, and of course to my son, who joined our family when I was in the middle of this book, and in many ways allowed me to see my work from a new perspective when I returned back to it. Thanks to him I had to rewrite (and I hope improve!) large chunks of it, teaching me that sometimes taking a step back is a big step forward. And while my daughter joined us when I was finishing the book, she was my motivation for pushing forward, and ultimately finishing the book, reminding me that a good book is a finished book. With all this assistance from so many amazing people, the only thing that is entirely my own are the errors, omissions, and weaknesses in the book.

Praise for *Translating Tourism*

"Sofia Malamatidou offers a comprehensive and insightful study of cross-linguistic differences in the promotion of tourist destinations and of the way these differences are managed in translation. This book's originality lies not only in its empirical nature but also in its multilingual perspective which includes six different language combinations, namely English (both American and British), French, Greek, and Russian. Indeed, tourist texts are based on a circular process involving cultural encounters: the way a destination is presented to readers is imbued with the socio-cultural features of that place or nation; at the same time, these texts are culturally filtered through the perceptions and expectations of reader-tourists. By providing a series of interesting analyses, Sofia Malamatidou shows that studying tourism discourse and its translation allows us to understand how cross-cultural communication works or should work and how 'we humans try to make sense of the world around us'.

A must read for experts and students in marketing and tourism studies and in translation studies, this book is an excellent contribution to the ongoing research on cross-cultural tourism discourse."

—Elena Manca, *Associate Professor of English Linguistics and Translation, University of Salento, Italy*

Contents

List of Figures

List of Tables

1

Introduction

1.1 Background

In recent years, public discussion around tourism has grown. Whether it is over-tourism in places like Barcelona and the effect this has on the local population, the unprecedented standstill of all tourism activities during the COVID-19 pandemic and how this affected economies worldwide, or more recently the effects of climate change on tourism patterns and what tourism might look like in the future, we have finally started to realise that tourism is not simply a leisurely, frivolous activity, but rather a collective human behaviour deeply ingrained in the fabric of modern life, influencing it and being influenced by it. This realisation has brought with it another one: tourism practices might need to change in the future, for example to create space for more sustainable tourism or rebrand cooler places as summer destinations. The discussion of these changes normally revolves around government policies and marketing campaigns, which are undoubtedly central if any change is to be designed and implemented.

S. Malamatidou, *Translating Tourism*, Palgrave Studies in Translating and Interpreting, https://doi.org/10.1007/978-3-031-49349-2_1

However, what is rarely, if at all discussed, is the important role that language (and by extension cross-linguistic variation) plays in shaping tourism practices and the key role that translation plays in tourism promotion. The language found in translated texts aiming to promote tourism is often criticised for its low quality and errors (such as a sign in Greece informing visitors that there are 'rooms to rent with atomic baths'—meaning ensuite), yet it is one of the most crucial tools through which international tourism is promoted (Sulaiman and Wilson 2019). During the process of writing this book, I have been involved in a number of conversations with national destination promoters of different countries and almost universally their reaction to the mention of the importance of language is one of complete unawareness. On one occasion, when asked about the importance of cross-linguistic differences in tourism promotion, a senior member of a tourism board replied that it is important to avoid any puns or idioms so as to facilitate translation. This sort of narrow understanding of the importance of language in tourism texts and the ways in which it is translated is widespread in the tourism industry. Even if there is some understanding of the varied cultural and linguistic markets that a destination aims to attract, this is very rarely reflected in the importance assigned to tourism language and how it is translated, which ultimately remains a purely semantic activity, that is, focusing almost exclusively on meaning. This results in pragmatic (e.g. how the promotional effect is achieved) and stylistic (e.g. what tone is being adopted) factors, among others, often being ignored. What is also ignored are cultural considerations that go beyond culture-specific items, such as how the readers and potential tourists might perceive and experience the world around them (which will be examined in this book). But it is not just the tourism and hospitality sector that pays little attention to tourism language and translation. Translation scholars have also not really engaged with research on tourism texts. It is revealing that while degrees, seminars, and workshops focusing on the translation of all sorts of subject fields (e.g. medical, legal, business, literary, etc.) are widely available at least in the Western world, similar training regarding tourism translation is hard to find (Durán Muñoz 2011), suggesting a conspicuous gap in translation research and by extension translation training.

This unawareness of the importance of language and translation in tourism-related texts needs to change, if it is to support destination promotion and facilitate the changes in tourism patterns we want to see in the future. This can only be achieved if we first gain a clear and critical understanding of the central role translation plays in destination promotion. What seems to be missing is an in-depth understanding of the crucial link between tourism translation, destination promotion, and how different people experience the world around them differently. Such an understanding would benefit not only translators who work with tourism texts but also destination promoters who want to attract a wider and more diverse audience.

Finally, there is perhaps no better example than tourism texts to reveal alternative worldviews, since these texts quite literally present the outside world to their readers. Tourism texts offer a window into how we perceive the world, which elements of it we prioritise, and how we relate to it through language. Therefore, tourism translation is much more than simply translating words. It is also about translating *worlds*. For translations of tourism texts to be successful, they need to present the worldview of the target culture while also accommodating the worldview of the reader. This connection between tourism texts and our perception of the world, and how it relates to destination promotion, has not been touched upon in previous studies focusing on the translation of tourism texts. As a result, our understanding of the conceptually complex mechanisms that allow different worldviews to be reconciled through translation remains fragmented and limited, and translations of tourism texts cannot fulfil their full potential to (re)create a destination image, a concept which will be explained in detail in Chapter 2, that is to be able to turn readers into visitors.

1.2 Aim

This book provides, for the first time, a large-scale empirical multilingual study of the cross-linguistic differences in the language of destination promotion, how they are negotiated in translation, and how, in doing so, the translated text reflects and reconciles different worldviews (that of the destination population and that of the reader-tourist). The concept of the reader-tourist is an important one and will be adopted throughout this book: the recipient of the tourism text is a reader in the first instance but has the potential to become a tourist, and it is this potential that tourism texts capitalise on.

Language, and by extension the analysis of linguistic differences, is understood in the present book in a broad sense as a means for encoding and describing our reality and for talking about the world around us. For example, societies that do not have the concept of Santa Claus will not have a word for it, while societies where politeness is a key concept will have many linguistic means for expressing it. Our linguistic choices reflect our perceived reality, and it is in this broad sense that I understand the link between language and cultural issues in this book. Also, 'linguistic' is used as a label to capture both semantic and pragmatic considerations (and potentially others, but these are not explored here), and appreciation of both, as will be shown in Chapter 3, is important when it comes to the translation of tourism texts. Therefore, I am also interested here in the communicative purpose of tourism texts, which is primarily promotional, and how this can be encoded in linguistic choices.

Data are taken from official tourism websites promoting tourism in the USA, UK, France, Greece, and Russia. Most previous studies have focused on English/Spanish or English/Italian—languages associated with a strong tourism industry—ignoring languages such as Greek, French, and Russian, whose respective countries were in the 10 most visited European destinations (UNWTO 2018), with France actually ranking first. Therefore, all languages examined here are associated with strong tourism promotion, either long-standing (e.g. Greece and France) or more recent (e.g. Russia). Although, during the course of writing this book, the situation regarding Russia as an international tourism destination has undoubtedly changed, data from this language are still examined, as they provide useful insights into how a society that is not typically 'Western' might employ tourism language for destination promotion and how it might perceive tourism more generally. Finally, by including two varieties of English, it is possible to examine whether there is indeed a universal version of English for tourism or whether different societies (i.e. American and British) approach destination promotion differently.

Official tourism websites have been deliberately chosen here, for both practical and scholarly reasons. On the one hand, these websites offer rich multilingual data, often being translated into many languages, while they also place much greater emphasis on the destination image than other types of tourism texts. On the other hand, they represent the official narrative of the destination concerning how it is to be perceived by both the local population and the rest of the world. As such, they are not only promotional, but to a large extent ideological, with a significant potential to shape perceptions about the destination. This particular ideological potential of tourism texts will be explored in the two case studies that form part of the present book.

Specifically, the book aims to examine:

- Differences in the linguistic means used to promote destinations in English (American and British), French, Greek, and Russian and what these specific linguistic choices can tell us about how these societies view the world around them.
- Whether and to what extent these differences are taken into account during translation, that is, whether translated texts stay close to source texts or meet the expectations of reader-tourists in the target language.
- What the implications are for destination promotion: how to convert readers into visitors and manage their expectations and how translation can become a field of active (re)construction of destination identity.

To achieve its aims, this book uses a corpus (i.e. a large electronic collection of texts) that I have specifically compiled for the purposes of this book, to examine linguistic patterns contributing to destination promotion. Specifically, the book examines the TrAIL (Tourism Across and In-between Languages) corpus, a 2 million-word corpus of both translated and non-translated tourism texts. Texts are taken from official tourism websites in English (American and British), French, Greek, and Russian and, where available, their translations into these languages (e.g. Greek into English, Greek into French, Greek into Russian, English into

French, Russian into English, etc.), promoting the UK and the USA, France, Greece, and Russia, respectively. A unique aspect of the book is that it is not limited to a specific language pair, but rather covers at least six different language combinations between English, French, Greek, and Russian, thus making this the first multilingual comprehensive study of its kind. Moreover, the book offers a rare insight into how English tourism texts are translated into other languages, a phenomenon typically neglected, despite the popularity of English-speaking destinations.

Interdisciplinarity is a distinctive element of the book, which goes beyond previous studies conducted in translation studies, and which combines research conducted in linguistics, translation studies, and tourism studies. Tourism texts offer translation studies a perfect test bed for the exploration of how different worldviews can be reconciled, while the detailed analysis of tourism texts educates tourism studies about the importance of cross-linguistic adaptation. Bringing these fields together, the book challenges traditional views of tourism translation as a field mainly preoccupied with semantic meaning and reveals the importance of tourism translation as an intellectual field that can offer valuable insight about how we perceive the world around us and how we relate to alternative worldviews.

The book focuses on the two elements that are indispensable for tourism to take place, which form the basis of the two case studies presented in the book. These are the 'things to do' and 'places to go' elements, or in other words the tourist and the destination. These case studies can be expanded into questions: What are the tourists encouraged to do when visiting a place? and How is physical space described and promoted? Each of these two case studies, the rationale for which is presented in more detail in Chapter 4, focuses on different linguistic features. Verbs are examined to reveal how tourists are expected to behave and engage with the destination and are therefore associated with the 'things to do' element, while adjectives are examined to help understand how physical space is described and presented to the reader-tourists and therefore are associated with the 'places to go' element. However, the discussion does not focus on a purely linguistic analysis resulting in a long list of verbs and adjectives preferred in one language or another. Rather the case studies are linked to key considerations in tourism, namely the concepts of tourism phase and tourist gaze. These concepts, which are discussed in detail in Chapters 5 and 8, respectively, offer an interpretive framework which supports the idea that tourism, and its associated elements, is not experienced or perceived in the same way across the world, but rather that important cross-cultural differences are expected. Briefly, tourism phase suggests that tourism as an activity is realised on a continuum from the sightseeing to the post-sightseeing phase, with different societies positioned at different points on this continuum. This is, therefore, a useful concept for the investigation of the 'things to do' element since it can provide an interpretive framework for why certain societies show preference for specific activities associated with tourism (e.g. simply moving through space vs exploring it). Tourist gaze suggests that the way we 'gaze' at the world around us, especially while acting as tourists, is subjective and differs depending on the cultural background, among other things, and serves as a useful framework for exploring how physical space is described and is, therefore, used for the examination of the 'where to go' element. Both concepts have as a starting point the (potential) tourist and their perceptions of tourism and its elements. They are considered to be complementary and aim to approach the same phenomenon from slightly different angles.

Relying on these concepts for the interpretation of results significantly enhances the interdisciplinarity element of the book and allows for a more comprehensive picture of the phenomena under investigation to be obtained. The focus in this book is on how these concepts (i.e. phases of tourism and tourism gaze) are manifested linguistically. By approaching linguistic choices made in tourism texts as meaningful and dynamic, shaping and being shaped by socio-cultural factors, it is possible to understand how these different elements come together to create a certain worldview, and also how translation navigates its way through different, and potentially competing, worldviews. This understanding is indispensable if we are to improve intercultural communication through translation, especially in the case of tourism, a field that offers a unique opportunity to experience other perspectives. Moreover, the book is relevant not only to practicing translators, but also to scholars interested in issues around tourism, promotion, and translation. From an intellectual perspective, the book foregrounds translation as a space where cross-linguistic differences around how we view the world around us can be negotiated. From a vocational perspective, the book will help translators of tourism texts reflect on their practice and ultimately produce translations that perform their aim more effectively. In turn, this can help transform the tourism industry and have a positive impact on the economies that rely on it. Finally, the book will also be of interest to destination promoters who aim to understand better the importance that language and translation play in tourism promotion.

1.3 Outline

This book aims to provide an account of tourism translation as a field where multiple and often competing worldviews can be negotiated and to showcase the significant potential it has to influence how the destination image is perceived and ultimately the promotional potential of the tourism message. It does not assume any prior knowledge of the topics discussed, each of which will be introduced in detail and with appropriate examples.

The book is divided into eleven chapters; this chapter and Chapter 11 introduce, and respectively, round off the volume while the remaining chapters are organised into three main parts: a theoretical, a methodological, and an analytical one.

Chapters 2 and 3 constitute the theoretical part of the book, which provides the background, reviewing research in tourism translation and emphasising its importance. It assumes that tourism translation is an interdisciplinary activity which requires an understanding of key concepts from a range of fields. Therefore, this part provides insights from the fields of marketing, tourism studies, as well as translation studies. Examining how exactly tourism promotion is achieved and the role that cross-cultural and cross-linguistic differences play in it allows us to understand the complex dynamics behind it, which are critical not only when non- translated tourism texts are produced, but more importantly in the case of this book, when tourism texts are translated into other languages, as they often are.

The discussion begins in Chapter 2 with an exploration of the language of promotion more generally and its relationship to the language of tourism, how exactly destination promotion is achieved and the importance of the Tourism Destination Image (TDI), as well as the main functions of the tourism text and how they are realised linguistically. The aim of this chapter is to clearly show that tourism discourse shares many of the typical features of advertising language and that its main aim is to promote a particular destination and create a positive image of it. Although this might appear as an obvious fact, there is surprisingly very little awareness around it from scholars and practitioners both in tourism studies and in translation studies. Chapter 3 focuses on translation studies more specifically and shows how this lack of awareness is manifested in the focus of extant research in the field. The vast majority of existing research is preoccupied with the semantic properties of texts, with pragmatic properties only rarely examined and not considered in light of the complex dynamics that need be taken into account for successful tourism texts to be produced. The theoretical part of the book closes with a discussion of the approach adopted in the two case studies which aims to uncover the missing link between the micro-level (i.e. linguistic features) and the macro-level (i.e. the socio-cultural considerations) analysis of tourism texts, and aims to understand how the socio-cultural preferences of how people perceive and experience the world around them are reflected in and realised through specific linguistic choices.

After establishing the importance of examining tourism texts in relation to the socio-cultural dynamics in which they have been produced, as well as the general mechanisms through which they turn readers into tourists, the book moves on, in Chapter 4, to the methodological part, which serves as a bridge between the theoretical and analytical parts, explaining how exactly the ideas discussed in Chapters 2 and 3 will be empirically investigated in the rest of the book. Since a corpus methodology focusing on official tourism websites is adopted, an overview of available data and how they have been selected and organised into a corpus is presented together with the corpus triangulation methodology that is employed for their analysis. A unique element of the book is that it provides a concrete methodology for the examination of adaptation patterns in translated texts which is informed by quantitative findings. It is worth noting that Chapter 4 provides an overview of the methodology employed in the two case studies, while the detailed methodological considerations of each case study are discussed in Chapters 5 and 8, respectively.

Chapters 5–10 constitute the final (and largest) part of the book, which has a strong analytical focus, providing details about the two case studies briefly presented earlier. The chapters for each case study follow the same pattern: an introduction to each case study and what it focuses on, as well as the specific methodology employed (Chapters 5 and 8), an analysis with detailed examples of non-translated texts (Chapters 6 and 9), and an analysis with examples of the adaptation patterns noted in translated texts (Chapters 7 and 10). For each case study, a clear link is provided between what element of tourism is being examined, how it is assumed to manifest itself linguistically, and what it reveals about how different societies perceive and experience tourism. Specifically, case study I (Chapters 5–7) focuses on tourist behaviour, that is how tourists are expected to behave and engage with the destination, which is examined through an analysis of verbs where the tourist is the actor or agent, which can provide information about the tourism phase that each society is experiencing. Case study II (Chapters 8–10) focuses on the destination, that is, how it is described and presented to the reader-tourists, which is examined through an analysis of adjectives modifying physical space. This provides information about the concept of tourist gaze and how it differs cross-linguistically. Overall, this book aims to make a significant contribution to challenging existing assumptions around the language of tourism and offer for the first time valuable insight into the important role that language plays not only in promoting tourism activities, but also in allowing people to talk about the world around them and their position in it.

References

Durán Muñoz, Isabel. 2011. "Tourist Translations as a Mediation Tool: Misunderstandings and Difficulties." *Cadernos de Traduçao* 1 (27): 29–49.

Sulaiman, M. Zain, and Rita Wilson. 2019. *Translation and Tourism.* Singapore: Springer.

UNWTO. 2018. "Tourism Highlights." https://www.e-unwto.org/doi/pdf/10.18111/9789284419876. Accessed 25 October 2023.

2

The Language of Tourism

2.1 The Language of Promotion

Although the language of tourism might not be immediately identified as an example of advertising, given how much lengthier, if anything, tourism texts are compared to typical advertisements, the main premise adopted in this book is that tourism texts have a strong promotional function, and for that reason they share many characteristics with advertising. Therefore, it is necessary to begin the study of the language of tourism by exploring the main principles of advertising, paying particular attention to the relationship between language and promotion.

Ideally, any study of advertising should focus on multimodality (Cook 2001; Gardner and Luchtenberg 2000), making use of the recent advances in digital humanities (e.g. see Burdick et al. 2012). However, this book is not about advertising more generally, but rather about tourism, and more specifically official tourism websites and how they are adapted into different language versions. While recognising the promotional power of images in destination promotion and acknowledging the research conducted on the topic (Francesconi 2014; Hallet and Kaplan-Weinger 2010), this book will only focus on language. The reason for this is that official tourism websites tend not to adapt the visual content

S. Malamatidou, *Translating Tourism*, Palgrave Studies in Translating and Interpreting, https://doi.org/10.1007/978-3-031-49349-2_2

according to the language version and thus do not offer sufficient data for comparison. Although the lack of adaption regarding images can certainly have its implications when it comes to the efficacy of the text and the extent to which visual images align with the idea of tourism that different reader-tourists might have, one of the key aims of this book is to understand what, if anything, changes during the translation process. Unlike visual images, where it is more straightforward to spot whether or not something has been changed or adapted, written texts require a much closer investigation, and differences are often analytically less conspicuous. It is precisely these less conspicuous differences that this book aims to uncover and, therefore, the focus here is on the verbal component of the tourism text, which is bound to differ across language versions. Moreover, the adaptation (or not) of visual images on tourism websites is typically the responsibility of marketing teams and translators often have little or no say in this process.

2.1.1 The Challenging Task of Advertising

Advertising is a paradoxical human activity: it is "everywhere but nowhere" (Cook 2001, 1). We are surrounded by human-made, often costly, advertising campaigns, and marketing departments around the world know well that advertising has an impact on consumer behaviour. And in order to make their advertising campaigns more noticeable, they engage in brand and image-building activities that are often outside what might be considered conventional advertising (i.e. the promotion of a specific product based on its features). A characteristic example is Red Bull's 2012 advertising campaign, when the company behind the well-known energy drink released a clip of an Austrian skydiver performing a jump approximately 39 kilometres into the stratosphere. The result was a 7% increase in sales in the six months following the stunt, raising the company's sales to $1.6 million (Bosomworth 2012). Red Bull's campaign was obviously aiming at attracting as much attention as possible.

In simple terms, the aim of advertising is to raise awareness, to grab consumers' attention, and to make them more favourably disposed to the advertised product. More specifically, the aim of advertising is to first persuade consumers to notice the advertisement among competing ones, and then to convince them of the superiority of the product over those of competitors (Bruthiaux 2000). According to Leech (1966), who was one of the first to examine Anglophone advertising, an advertisement, in order to be successful, needs to fulfil a number of functions: it needs to attract attention, be readable (at a time when advertising on audio-visual media was fairly new), memorable, and have selling power. Similarly, Bolen (1984) argues that advertisements need to be believable, simple, and readable. The message communicated also needs to be relevant to the target audience (Bulmer and Buchanan-Oliver 2006), which can, however, be quite diverse, if we consider factors such as income, educational background, and personal circumstances. The situation is bound to become even more complicated when different cultural groups are being targeted, and many advertising campaigns are typically aimed at a diverse audience (similar to many translation briefs), which to a large extent is unidentifiable and can only be guessed at by marketing departments. In addition to this, advertising is not the creation of a single individual, rather the result of a collaborative project (from the manufacturer to the agency and the creative team), and, for each individual involved, the perceived focus of the specific advertisement might be different (Cook 2001). This is why Bruthiaux (2000, 299) views advertising as being performed in "a communicative fog".

2.1.2 Attention-Seeking Devices

To achieve its aims, advertising relies on a number of attention-seeking devices or means of communication, which Cook (2001) refers to as modes: music, image, and language. Although, as explained above, the focus in this book is on written language only, the other two modes are briefly presented here to highlight the complexity of advertising and, importantly, the key role that cultural differences can play in it.

For multimodal advertising campaigns, such as those produced for TV or radio, music can play a vital role in shaping consumption patterns. Taylor (2012) argues that, because music can affect us emotionally, it has become a powerful tool for advertisers and has significantly contributed towards making us the consumers we are today. Similarly, Oakes (2007) concludes that music in advertising can enhance purchase intent, brand attitude, recall facilitation, and affective response. Research in psychology (Zander 2006; MacInnis and Park 1991; Hung 2000, 2001; Mattila and Wirtz 2001) has also indicated that it is not just the presence of music that creates an effect in advertising, but rather that different musical styles (e.g. rock or classical song) can generate different impressions of the brand. It is often argued that music, and for that matter the right type of music, can affect consumer response in ways that words cannot, highlighting the power of music over other modes of communication. However, like language, music has been found to be culture-dependent (Scott 1990), which is not surprising if we consider the diversity of musical styles, some of which tend to originate and be associated with specific parts of the world (e.g. Latin music), not to mention the enormous diversity of folk music. Murray and Murray (1996) were some of the first to recognise the non-universality of music in advertising and argue that music needs to be understood in a particular cultural and social context. The use of music in advertising subtly communicates cultural values and reflects cultural differences, which, they argue, need to be carefully harnessed by marketers. Therefore, knowledge of the cultural associations of musical forms can help predict how consumers might react to certain musical selections for an advertisement (Alpert et al. 2005). Just like musical tastes, the effects of music are not universal.

Similar to music, pictures can also be very powerful modes of persuasion, and Bulmer and Bucharan-Oliver argue that visuals are "at least as powerful as words" (2006, 58), while research suggests that pictures can attract attention (Pieters and Wedel 2004) and generate emotional responses (Chowdhury et al. 2011). More specifically, attractive visuals, such as beautiful scenery, might create a positive predisposition and improve the attitude towards a brand (Pieters and Wedel 2004; Kim et al. 1998; Xue 2014). Additionally, visuals might also act as cues, grabbing readers' attention and making them engage with the (textual) message (Petty and Cacioppo 1991; Belch and Belch 2001). Although, in the past, research (e.g. Moriarty and Duncan 1991) in advertising has perceived images as universal and visual communication as more straightforward than verbal communication, recent research provides strong evidence that visuals are extremely complex and rich (Phillips and McQuarrie 2004). A single picture, as used in traditional magazine advertising, can represent a wide range of "concepts, abstractions, actions, metaphors and modifiers" (Scott 1994, 253). If we combine this with the widely accepted premise that images are "culturally bound, symbolic and subject to the viewer's interpretation" (Bulmer and Buchanan-Oliver 2006, 51), it becomes clear that a 'visual Esperanto' (Callow and Schiffman 1999) cannot exist in advertising. Importantly, research has shown that, in a cross-cultural setting, the same visual content will elicit different interpretations (Callow and Schiffman 1999; Bulmer and Buchanan-Oliver 2006). Therefore, although concepts behind images might be shared by different cultural groups, the way they can be communicated visually is often not, and differences in interpretations might even result in the failure of an advertisement if visual material is not adapted accordingly.

Finally, since persuasion is one of the main aims of advertising, a significant amount of research around the language of advertising has focused on the rhetorical devices used (Crompton and McAlea 2000; McQuarrie and Mick 1996; van Mulken 2003). A rhetorical device is an "expression [that] deviates from expectation, [but] (…) is not rejected as nonsensical or faulty" (McQuarrie and Mick 1996, 425); elsewhere, it is defined as a form of artful deviation (Corbett 1990). A characteristic of the language of advertising, according to Berlyne (1971), is incongruity, since any deviation from typical language use is likely to attract attention, which is important if we consider that consumers often have the option of engaging or not with an advertisement (e.g. in print media). What is more, the more artful this incongruity is, the more positive the effect of the advertisement is likely to be (McQuarrie and Mick 1996). The main challenge for advertisers is to "play with, push, and extend the limits of language, breaking rules, but nevertheless retaining coherence" (Gardner and Luchtenberg 2000, 1807).

Based on the main premise of rhetoric, according to which a proposition can be expressed in many different ways, some more effective than others in creating the desired response in the audience, studies examining the language used in advertising focus not on what is expressed, but on how it is expressed. In that sense, advertising is often compared to literature and poetry (Cook 2001; Stern 1988) as both seek to affect their audience through the use of evocative language. Examining Anglophone advertising, Leech (1966) was one of the first to highlight the rich repertoire of linguistic devices used in advertising and identified a number of linguistic features which he associates with the different functions of advertising. Specifically, according to Leech, advertisements attract attention by using illustration, elliptical constructions, metaphors, and paradoxes. Some linguistic devices are related to memorability, such as alliteration, rhyme, and onomatopoeia. Readability is increased through the use of a simple and colloquial style, while selling power is associated with the use of imperatives, superlatives, inversions, and parallelisms. More recently, Myers (1994) also identified assonance, homophones, puns, and parody, while Brierley (1995) also lists repetition, similes, paradox, omission, and ambiguity. Further linguistic devices such as metonymy have been identified by Cook (2001), while Fuertes-Olivera et al. (2001) focused on the use of personal markers, hedges, emphatics, endophoric markers, and evidentials. Without going into too much detail about these linguistic devices, it is evident that to communicate their message effectively, advertisements need to extensively manipulate language and can choose from a range of tools to help them achieve their aim.

2.1.3 Linguistic Adaptation in Advertising

At the rise of modern globalisation, there was a widely held assumption that consumer attitudes and marketing techniques need to converge regarding the textual message. This meant that Western—typically North American—middle-class marketing concepts and practices were imposed on the rest of the world, with little attention paid to the importance of language and cultural differences. Early research into marketing (e.g. Bagozzi 1975; Kotler and Levy 1969) does not recognise the importance of cultural differences in international marketing, arguing that globalisation is expected to bring worldwide convergence, even though there was still very little empirical evidence that we can talk about a true homogenisation of tastes or the existence of universal consumer segments. Usunier and Lee (2005) attribute this fixation on homogeneity to the status of English as a lingua franca: many native English speakers are said to naively assume that there is no reason to learn foreign languages, which makes them less sensitive to the importance of cross-cultural communication. They add that international businesses need to develop "an awareness of what language differences imply" (ibid., 394), which requires a complex understanding of intercultural communication and goes far beyond simply learning grammar and vocabulary. The language in which a message is communicated will shape how it needs to be communicated to be successful, since it is language, not the product advertised, that defines the rules of the game.

There are different ways in which differences in language and culture can affect an advertising strategy focusing specifically on the verbal text. One such way is in terms of the overall communication style, that is, whether it is mainly persuasive, informative, or oneiric (dream-oriented) (Usunier and Lee 2005). Early research (Stern and Resnik 1991; Chiou 2002; Cutler et al. 1992; Lin 2001; Usunier 2006; Zandpour et al. 1992; Martenson 1987) suggests that different cultures will show a preference for different communication styles. Abernethy and Franke (1996) conducted a meta-analysis of 59 studies about information content and concluded that developed countries prefer a more informative style of advertising compared to developing countries. However, variation can be observed within the developed world, and even between neighbouring countries, with Germans found to prefer informative advertising, while the French rely more on the oneiric style (Usunier and Lee 2005). Although many factors can explain this difference in preferences, such as the type of market and competitive environment or legislation around advertising, Usunier and Lee argue that "culture is the most important factor in explaining how much and what kind of information can be found in advertising" (ibid., 415). But it is not just about the information that is *found* in advertising. Culture will also affect how much and what kind of information is *sought* in advertising. For instance, it has been found that Asian consumers rely on a decision-making process that is "more intuitive, less scientifically valid, more mixed with direct implementation" compared to American consumers (Usunier 1996).

Therefore, advertising strategies need to be adapted depending on the recipients, and this adaptation is to a large extent linguistic, since, at least in the case of verbally based advertising, culture can be expressed though linguistic choices.

Linguistic adaptation is not only related to the overall communication style—more subtle linguistic differences might also play a role. For instance, Tixier (1992) argues that for an advertisement to be effective in a French context, half of the words need to be nouns or verbs, long words (i.e. more than three syllables) should represent only a very small proportion (i.e. no more than 10%), and sentences should be between 10 and 13 words. When it comes to adapting an advertisement for a different linguistic audience, such detailed instructions reflect the importance of not simply translating the meaning of words, but reorganising the entire message so that it meets the expectations and preferences of the target audience. But even translating the meaning of words can be a complex process that requires knowledge of cultural differences. An example provided by Kroeber-Riel (1992), which is relevant to the language of tourism, has to do with the translation of the English word *quiet* into German (*ruhig*) and French (*tranquille*). Kroeber-Riel found that when reading the German word, the majority of German consumers (41%) think of a forest, with sleep coming right after (35%), other pictorial associations being church (20%), cemetery (13%), and bed (8%). French consumers think of different and more varied things (13% think of countryside, 11% of forest, 9% of house, 5% of library with smaller percentages for the rest). In fact, very little congruency was found between the two groups. Linguistic adaptation is not only about denotative meaning, but importantly about finding the words and expressions that will generate the right associations in the mind of the consumers, in line with the communicative style and the linguistic preferences of advertising in that language.

The importance of language in advertising has also been highlighted in studies focusing on how bilinguals process information in advertising (Noriega and Blair 2008). The main focus has been on whether advertising directed at bilingual minorities is more effective if presented in the bilinguals' native language than the country's dominant language. Perhaps the most well- known study is that conducted by Luna and

Peracchio (1999, 2001) who conducted experiments with Spanish–English 'bilinguals' in the USA. They concluded that advertisements that relied on the 'bilinguals'' 'native' language are more successful, simply because they were easier to process. They later extended their research (2002, 2005a, b) to examine the effective response associated with the use of the 'bilinguals'' 'native' language, arguing that some words activate a stronger emotional response when presented in the 'native' rather than the 'dominant' language. Their research offers evidence that it might be more effective to use 'native' (i.e. minority) languages when advertising to members of such groups. However, it is not simply about the language used, but importantly also about what is being advertised. Wyer (2002) argues that advertisements that emphasise collectivist goals are likely to be better received by Asian bilinguals when they are conveyed in their native language, whereas if they emphasise individuality, then they will be better perceived by the same group if conveyed in English, since Asian cultures score higher in terms of collectivism, while Anglophone countries tend to be associated with individuality according to Hofstede's (2001) cultural dimensions, although it must be noted that Hofstede's work has often been challenged on various grounds (Dorfman and Howell 1988; Newman and Nollen 1996; Schwartz 1999).

The power of language in advertising has also been explored in studies focusing on the use of both foreign and local languages in advertisements (Bishop and Peterson 2010; Caroll and Luna 2011; Khan 2014; Luna et al. 2005; Luna and Peracchio 2001, 2002, 2005a, b; Vettorel 2013). This phenomenon of inserting linguistic elements (e.g. words or phrases) of one language into another is referred to as codeswitching (Grosjean 1982), and is also found in tourism texts, where it has been referred to as 'languaging' (Dann 1996). This has become a powerful technique in advertising, which can be explained by the Markedness Model (Myers-Scotton 1993), according to which codeswitching can carry additional connotations in certain social contexts. For instance, English is very often inserted into non-Anglophone advertising as "an attention getter and can serve as a mood enhancer" (Martin 2008, 49). From a linguistic point of view, foreign words are not used for their meaning, but rather for their "ethnocultural associations" (Hornikx et al. 2013, 153). The use of foreign words activates language schemas (Luna and Peracchio 2005b) which include the individual's perceptions about language, its social and cultural meaning, as well as attitudes, concepts, and beliefs around language. In simple terms, if the foreign language evokes positive associations, then the advertised product will be perceived positively (Baumgardner 2006; Leclerc et al. 1994; Luna and Peracchio 2005a). This can have important implications for the persuasiveness of the message. Naturally, advertisers need to be aware of the target market's attitude towards the foreign elements (Ahn et al. 2016) and the foreign language used needs to be congruent with the products advertised (Domzal et al. 1995; Kelly-Holmes 2000, 2005; Ray et al. 1991). For example, French words used in an Anglophone advertisement for chocolates aim at highlighting their luxurious aspect, while German words in Anglophone car advertising aim at stressing its reliability and high quality.

The aim of the discussion around the power of language in this section is to highlight the complex role that it can play in advertising, without denying the strong effect that sounds and images exert. From selecting the right language, to identifying the appropriate communication style, and then the specific words that are more likely to give rise to desired associations, while at the same time considering who is being targeted and what is being advertised, language operates at many levels simultaneously. Not only is it not possible to argue for worldwide convergence and homogeneity, but because of considerable divergence and pluralism, we still know relatively little about the effect that language has on persuasiveness. If creating an advertisement is a challenging task, adapting it for a new linguistic audience poses further challenges, and advertisers are expected to be sensitive to cultural differences and carefully select the linguistic means for their messages. Considering official tourism websites, which are typically verbally rich, it is clear that linguistic adaptation is bound to be significant. The next section will explain why cultural differences need to be taken into account when adapting a tourism text to a new audience.

2.2 Destination Promotion

2.2.1 Tourism Destination Image (TDI)

When it comes to tourism promotion, which is associated with a lucrative yet highly competitive market, Morgan et al. (2002, 4) argue that "places are potentially the world's biggest tourism brands", while the World Tourism Organization (2001) predicted that in the next century tourism destinations would become fashion accessories, helping travellers set themselves apart from other tourists, something which has certainly manifested itself through the use of social media. As a result, destination branding is bound to become even more critical in the future. Although similarities can be observed between the branding of goods and destinations, with some scholars arguing that the two can be branded in the same way (De Chernatony and McDonald 1992; Pritchard and Morgan 1998), branding destinations is bound to be more difficult, as a result of the complexity (Smith 1994) and multidimensionality (Gartner 1989) of tourism products. A fundamental difference between destination promotion and the mainstream advertising of goods is that, although they require a significant financial investment, holidays are intangible products which cannot be evaluated beforehand. Destination selection is, therefore, largely based on invisible pre-visit elements and subjective impressions. In a highly competitive tourism market, destinations need to create a unique identity, which will allow them to differentiate themselves from their competitors. What becomes important is the image that the tourism product creates in the consumers' minds, rather than its actual characteristics (Morgan and Pritchard 2002), since "perceptions, rather than reality, are what motivate consumers" (Guthrie and Gale 1991, 555). In the post-modern age, image, whether real or mental, is superior to reality. It is important to clarify here that Tourism Destination Image (TDI) is indeed a conceptual construct that exists in the minds of potential tourists, rather than a visual image, and is created as a result of different factors, as will be explained later in this section. Some of these factors are related to the image of a destination as created by the tourism text which affects the perceptions that reader-tourists form when reading such texts.

The importance of visual imagery in tourism advertising has been identified as early as the 1970s by Gunn (1972), Mayo (1973), and Hunt (1975), when it was established that tourism imagery is not singular in nature, but plural in meaning and ideology. In more recent years, the Tourism Destination Image (TDI) is widely recognised as the most important element affecting tourist behaviour and turning recipients of tourism promotion into visitors (Chen and Tsai 2007; Nadeau et al. 2008; Govers et al. 2007; Tasci and Gartner 2007). TDI has also been found to play an important role throughout the entire tourist experience (Galí and Donnaire 2005): during pre-visiting when a destination is selected among many; during visiting when expectations are compared to reality; and during post-visiting when decisions are made about revisiting a destination and/or recommending it to others.

As a result of the volume of research on TDI and its complex and elusive nature, there are as many definitions as there are researchers (see, for example, Agapito et al. 2013; Bigné et al. 2001; Chon 1990; Crompton 1979; Gallarza et al. 2002; White 2004). Such definitions try to capture the subjective nature of TDI, defining it as the 'impression', 'perception', and 'interpretation' of a destination, or the 'beliefs' and 'ideas' about a place. However, 'attitude' is argued to be the most suitable way to represent how a destination is understood and how it might affect the decision-making process (Frias et al. 2011; White 2004), following the definition provided by Baloglu and McCleary (1999, 870) according to which a TDI is "an attitudinal construct consisting of an individual's mental representation of knowledge (beliefs), feelings and global impression about an object or destination".

Understanding how exactly a TDI is created, as well as the elements of its plurality, is of paramount importance, since tourism can profoundly shape the country's economic destiny, as well as its sense of identity (Hallet and Kaplan-Weinger 2010). TDI is naturally closely linked to destination branding, since one aspect of such branding is the imagery promoted. This needs to rely on originality and difference, since the same imagery is often used by different destinations to attract visitors, but at the same time it needs to be sustainable, believable, and relevant (Morgan and Pritchard 2002). Although not mentioned in the literature on TDI and destination branding, translation plays a key role in communicating the destination brand globally and helps promote a certain imagery, considering that consumers quite frequently do not have access to the language in which the tourism text was originally written. However, the majority of national tourism organisations have limited budgets, yet they have to market globally, which often means that translation is not a priority within this overall limited budget.

It is important to stress that a TDI may or may not correspond to reality (Andreu et al. 2000; Bennett 1995), and, considering the key role that it plays in destination promotion, it might be assumed that the reality of tourism is less important for destination promoters than its expressive representations. Govers et al. 2007 warns about the dangers of overreliance on imagery though, which can result in the creation of a tourism product which is not in line with destination identity. The relationship between the perceived TDI and the actual tourism experience has been defined as "evaluative coherence" by Chon (1990), who argues that the higher the evaluative coherence, the higher the satisfaction level, which is one of the determining elements in destination evaluation (Bigné and Zorio 1989). If evaluation coherence is low, this means that tourist expectations, as shaped by an often unrealistic TDI, have not been met during the tourism experience, which will have a negative effect on repeat visits and the dissatisfaction is likely to be communicated to other individuals (Gitelson and Kerstetter 1994). This scenario is not infrequent, as what is delivered in terms of tourism experience tends to be different from the tourist's expectations, which are often exaggerated by tourism promotion.

2.2.2 Destination Image Formation

Definitions of TDI, such as the one by Baloglu and McCleary (1999) provided earlier, try to identify some of the components that contribute to destination image formation, particularly highlighting the importance of psychological factors. However, given the key role that TDI plays in informing the destination selection process by reader-tourists, it is necessary to better understand how a destination is selected among many and, hence, examine in more detail how exactly a TDI is formed, or as eloquently put by Brokaw (1990, 32) "[b]efore image can be used to influence behaviour, it is important to understand what influences image". A number of models have been proposed over the years focusing on destination image formation. It is worth highlighting that these models do not compete with each other but have developed based on each other in an integrated manner, with each new model building on the previous one. This has resulted in significant overlap, with differences observed mainly regarding terminology.

Gartner (1993) was the first to propose that TDIs are formed by three "distinctly different but hierarchically interrelated" components—cognitive, affective, and conative—and many researchers have since supported this classification (Dann 1996; Pike and Ryan 2004; Tasci and Gartner 2007; Tasci et al. 2007). The **cognitive (also known as intellectual or perceptual) component** relates to the reader-tourist's knowledge and beliefs about the attributes of a destination. It derives typically from facts (e.g. geographical location), whether true or perceived as being true. This creates an internally accepted picture of the destination; it refers to what we think about a destination. For instance, the knowledge that Greece is situated in southern Europe will give rise to an image of a hot and sunny summer destination. The **affective component** is guided by the feelings that we have about a destination. This is strongly activated during the evaluation process, that is, when selecting from a number of possible destinations, and is tightly linked to the motivation of travel. For instance, having positive feelings about Greece will increase the chances of it being selected as a travel destination. Finally, the **conative component** refers to the action of selecting a destination (it may also refer to revisiting a destination and/or recommending it to others) and is therefore the 'outcome' of the other two components. For instance, one might select Greece instead of Italy as a travel destination. Given the pragmatic nature of this last component, it is often seen as distinct from the other two, and that is why it is missing from more recent models (e.g. Baloglu and McCleary 1999; Beerli and Martín 2004). These components combined give rise to the **global component**, which is the overall image of a destination as a result of both cognitive and affective considerations. This model is inspired by Boulding's (1956) research on individual behaviour and social dynamics, according to which an image consists of three elements: what one knows and thinks about an object (cognitive), how one feels about it (affective), and how one acts using this information (conative).

Focusing on the cognitive and affective components, there exists a hierarchical relationship between the two (Baloglu and McCleary 1999). Specifically, the cognitive component is said to be activated first, followed by the affective component, but the two are not independent—affect is informed to a large extent by cognition. Understanding how cognitive and affective components affect TDIs can be used to predict tourists' reaction to a destination. It has also been proposed that the affective component has a stronger impact on the global component than the cognitive one (Cai et al. 2004; Kim and Yoon 2003; Konecnik and Gartner 2007; Pike and Ryan 2004; Tasci et al. 2007). This is not surprising if we consider the importance of emotional response in advertising, and research has shown that emotions might be better predictors of behaviour than perception (Yu and Dean 2001). For this reason, Sirgy and Su (2000) argue in favour of avoiding informational advertising when addressing people who have not visited a destination before. This provides further support to the idea that the main aim of a tourism text is to promote rather than to inform, even though at first glance such texts might appear as largely informational (see Chapter 3 for more detailed discussion).

Research on TDI formation might appear as elusive, considering the many, often subjective, factors involved, but it is nevertheless significant as it helps us to better understand how exactly a destination might be chosen among many. This is because, as mentioned earlier, potential tourists are purchasing an intangible product which cannot be experienced until after the purchase. Unlike typical products, tourism involves many more risks, which can be monetary, functional, physical, social, and psychological (Jacobsen and Munar 2012). Risk reduction is necessary in the case of tourism, especially if it is the first time visiting a place, which can be achieved by engaging in extensive information searches before purchasing (e.g. Gursoy and McCleary 2004).

2.2.3 Factors Affecting Tourism Destination Image

According to Gartner, destination image formation is driven by several 'agents', which he defines as "force(s) producing a specific result" (1993, 197); in the case of tourism, the desired result is to create an attractive TDI. Essentially, 'agents' refer to the different types of information that individuals are exposed to, which can affect the destination selection process. These include information sources (e.g. brochures, TV, and radio ads), word-of-mouth, as well as previous personal experiences of visiting a destination. More recent models (Agapito et al. 2013; Baloglu and McCleary 1999; Beerli and Martín 2004) also recognise the importance of psychological and social criteria and introduce (inter)personal factors to the model. Beerli and Martín's (2004) model is schematically presented here due to its greater semantic transparency (Fig. 2.1).

According to Beerli and Martín, a TDI is affected by both information sources and personal factors. Information sources can be considered 'pull' factors attracting an individual to a particular destination, while personal factors can be considered 'push' factors' consisting of the variables that predispose an individual to travel more generally.

2.2.3.1 Information Sources

Information sources are divided into primary and secondary, and can affect both cognitive and affective components. **Primary information sources** relate to information acquired through personal experience, for

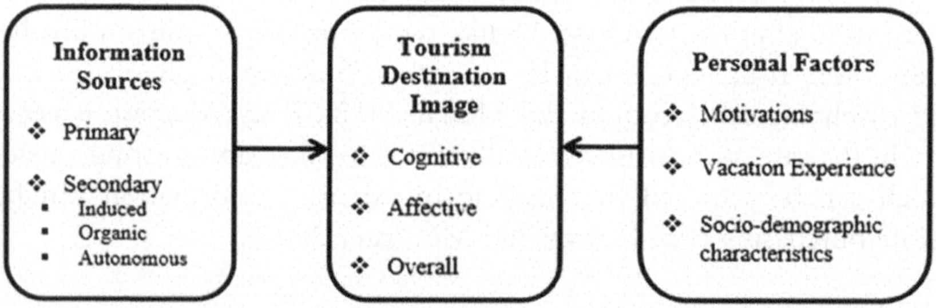

Fig. 2.1 Tourism Destination Image formation

example by visiting a destination. Factors such as the intensity of visit (e.g. time spent exploring the area) will naturally affect the TDI. Subsequently, using the TDI created during the first visit as a source informing return to a destination can have a strong impact on destination selection.

Secondary information sources are divided into induced, organic, and autonomous. Their main function, according to Mansfield (1992), is to help create an image of the destination, to minimise the risk involved in destination selection, and help justify destination choice. **Induced information sources** originate from marketing efforts of destination promoters and can be either overt or covert. **Overt induced sources** include traditional forms of advertising (e.g. TV, radio, brochures, etc.) that are associated with the official marketing efforts of destination promoters (e.g. tourism boards) or produced by organisations whose aim is to promote a particular destination, even though they might not be associated with that particular area (e.g. destination information on a tour operator website). **Covert induced sources** include second-party endorsements either by using someone recognisable (e.g. a celebrity in a tourism advertising campaign) or through written reports, such as newspaper articles (e.g. travel writing), which although appearing unbiased, in reality, are often guided by overt induced sources. Unlike induced sources, **organic sources** are relatively unbiased and independent of the destination and do not have control over destination. They form a large category which includes word-of-mouth, or more recently 'word-of-mouse' (through digital forms), accounts of visiting a destination, whether solicited or unsolicited. Finally, **autonomous sources** include independently produced news and popular culture, such as documentaries, news stories, and movies. These might involve reports commenting on the political situation or economic conditions of a destination, technological advancements, or social changes, either positive or negative. The focus is not on promoting a destination, but images of that destination might be present (e.g. a news report about an earthquake in south Asia or a film shot in Paris).

Earlier research into the impact of information sources on TDI suggested that the Internet did not play a significant role in the formation of the cognitive component (Beerli and Martín 2004). However, as potential tourists became more IT-skilled and tourism professionals found better ways to exploit the possibilities offered by web platforms, the importance of the Internet for destination selection has increased. In recent years, the entire tourism experience has been transformed by advances in IT (Buhalis and Law 2008), making it easier than ever to publish online content, while a significant rise has been noted in online bookings and ecommerce in tourism (Fesenmaier and Cook 2009). Nowadays, the vast majority of tourism enterprises, including national tourism organisations, communicate with prospective tourists through the Internet. Further support to this argument is provided by the most recent Eurobarometer (2021) survey on the attitudes of Europeans towards tourism, which identifies different online sources (i.e. websites and social media) as some of the most important sources of information. As a result, the majority of recent studies agree that the Internet is an important information source when selecting a destination, playing a crucial role in TDI formation (Bastida and Huan 2014; Fernández Cavia et al. 2014; Kim and Lehto 2013; Phau et al. 2010; Rodríguez-Molina et al. 2015), while some (Anderson 2006; Hwang et al. 2006; Kim and Fesenmaier 2008) argue that it is the most persuasive means available to influence destination selection. In the last decade, the rise of user-generated content on social media and in the form of travel blogs and reviews has also attracted attention as a new way of managing TDI (Christodoulides and Jevons 2011; Christodoulides et al. 2012; Huang et al. 2010).

The proliferation of different web platforms means that they can follow the classification of traditional offline information sources, namely induced (mostly Web 1.0, e.g. official tourism websites), organic (mostly Web 2.0, e.g. social media), and autonomous (mostly Web 1.0, e.g. news items). Focusing specifically on official tourism websites (i.e. websites produced by a government or an official tourism body), which form the focus of this book as they are translated into multiple languages, these have been claimed to be "one of the most important tools that destinations have to disseminate or project their image" (Marine-Roig and Clavé 2016, 255). Although in the past (Gartner 1993) induced forms of advertising have been perceived as least influential further research into official tourism websites suggests the opposite, arguing in favour of the influence they can exert on potential tourists (Bastida and Huan 2014) and even their capacity to change a TDI (Dragova et al. 2014). Hallet and Kaplan-Weinger (2010) stress the importance of official tourism websites, which they argue serve as important mediators between potential tourists and destinations. Through this mediation, visitors to the websites are engaged in the construction of self as tourists and of locale as a tourist site. Finally, according to the detailed study conducted by Llodrà-Riera et al. (2015), who investigated the information sources that potential visitors to Mallorca consult, the Internet, as a general category, was found to be the most useful source, followed by organic sources, such as friends and family. In terms of web platforms, most useful were considered to be search engines and online maps, followed by webpages created by official tourism websites.

Related to this is research comparing official government websites to other tourism-related websites (e.g. tour operators, travel agents, travel guides, etc.) which concludes that official tourism websites place greater emphasis on destination image than more marketing-focused websites (Choi et al. 2007; Govers and Go 2005). Based on this, Lepp et al. (2011, 676) argue that such websites are the "ideal starting place for exploring the Internet's potential to induce a favourable destination image". This is further supported by a study conducted by Mohammed (2004) who found that the vast majority of small developing countries had official government tourism websites, and a significant proportion of these websites focused on promotion, that is, on creating a positive TDI through self-presentation.

As this is still a nascent area of investigation, more research is needed into the role that the Internet plays as an induced source and how exactly information is presented, since typically a considerable budget is spent on the development of official tourism websites. For instance, Loda et al. (2009) argue that destination promoters should focus on presenting basic information in a credible way, instead of spending energy to develop novel website elements. If basic information does not appear to be credible, this can have a negative effect on TDI and might encourage potential visitors to seek information about the destination elsewhere (online or offline). Similarly, Rodriguez-Molina et al. (2015) conclude that a TDI is more positive when the website includes emotional triggers and advocate the use of more emotive messages. Finally, since the Internet offers each individual tourism supplier the opportunity to easily present themselves and communicate with potential consumers, there must exist a coherent strategy to develop and promote a destination image (Palmer 2002), which is why it is important to gain a deeper understanding of how a TDI is formed, not least linguistically.

2.2.3.2 Personal Factors

Personal factors also have a strong effect on TDI, mostly through the affective component. They consist of the psychological (e.g. motivations, preferences, personality, lifestyle), the socio-demographic characteristics of individuals (e.g. age, social class, and country of origin), and previous vacation experiences, since past experiences can affect the interpretation of present situations (Schreyer et al. 1984). Even though we might be exposed to the same external information sources, insofar as our cognitive structures are different, we will form different images of a destination. Ultimately, what convinces potential tourists to visit a place is whether they have "empathy with the destination and its values" (Morgan and Pritchard 2002, 12). For instance, the same potential destination will be viewed very differently by a teenager and a retired person, depending on whether they live in a big city or a rural area, whether they enjoy outdoors activities, and so on.

It is within the category of personal factors that we find the most interesting elements for cross-linguistic research. Specifically, one of the socio-demographic characteristics typically identified is the country of origin, with various studies suggesting that it can affect the perceived destination image (Baloglu and Brinberg, 1997; Baloglu and McCleary, 1999; Beerli and Martín, 2004; Chen and Kerstetter, 1999; Walmsley and Jenkins 1993). In fact, Beerli and Martín (2004) have identified the country of origin as the socio-demographic characteristic that has the greatest influence on both first-time and returning visitors. The reason for this is that our cultural background affects our cognitive structures, allowing us to interpret the world differently and create different associations with destinations.

Country of origin is naturally tightly linked to cultural background. Without going into an in-depth discussion of culture, national culture is understood here as "the collective programming of the mind which distinguishes the members of one human group from another" (Hofstede 2001, 21). The limitations of associating country and nation with culture aside, the notion of national culture can be useful in distinguishing among different groups (Hsieh and Tsai 2009), and it has been used to explain many of the patterns found in tourist preferences and behaviours (You et al. 2001; Reisinger and Turner 1998). Some studies even suggest that culture might play a role in the information searches carried out (Gursoy and Chen 2000; Gursoy and Umbreit 2004; Money and Crotts 2003; Uysal et al. 1990) with some cultures, for example, showing a stronger preference for the Internet compared to travel agents than others. In summary, previous studies agree that culture plays a key role in the overall destination selection process.

Since destinations are not singular units, but rather have the potential of multiple interpretations based on the characteristics of the perceiver, destination promoters need to decide not only which elements would work best in attracting visitors (Govers and Go 2005), but importantly how they need to be adjusted, not least in terms of the linguistic and cultural background of potential visitors. This is in line with advertising research discussed earlier stressing the importance of culture and language for promotion. What this means is that, no matter how convenient, a single, consistent image of a destination is unlikely to work for destination marketing purposes. This echoes the debate around globalisation and localisation. Although early research into tourism assumed that there is an underlying universality behind tourism behaviour—similar to the assumption that consumer attitudes converge globally—in reality there is a significant "segmentation of international markets" (MacKay and Fesenmaier 2000, 417), which results in distinctive locales defined as "the combination of a sociocultural region and a language" (Jiménez-Crespo 2013, 12). The problem with a standardised approach to destination promotion is that the image promoted can be interpreted differently in different cultures and, notably, not how the producer intended it. It is important that destination promoters segment the tourism market (Kozak 2002), recognise what each market is attracted to, and adapt their communication strategies accordingly (Beerli and Martín 2004). However, it is also possible that several cultures interpret the destination in a similar way and a destination can generate certain "internationally shared meanings" (Therkelsen 2003, 135), that is, a global image of a destination. This does not mean that a global approach is more effective, but rather that cultures, as might be expected, share certain characteristics. Subsequently, Therkelsen advocates for a glocal approach to tourism destination promotion, which "combines global and culture specific meaning structures and thereby holds the potential of creating a world-wide recognizable image which simultaneously has a specific appeal to local markets" (2003, 136). In other words, there is no need for a completely new approach each time a new audience is being addressed, rather a global approach with locally adapted features.

Recognising the importance of adaptation, the next question that needs to be addressed is how this adaptation is to be achieved in terms of the written text, which is the focus of this book. Going back to the key role that personal factors play in TDI formation, it is perhaps convenient to associate country of origin, and by extension culture, with language, and in many cases this association might prove helpful, for example when adapting a British tourism text for a Japanese audience. However, countries, or locales, should not be confused with languages, as the same language might be shared by a number of locales, with different sociocultural characteristics. A typical example is English, which is spoken as a first language in the UK, the USA, Australia, and many other countries of the world. In addition to this, many tourism websites often adapt their pages into a kind of 'international English', whatever that might mean, without a clear association with a specific locale. The same can be observed for other widely spoken languages, such as French and Spanish.

Ideally, a tourism text needs to be adapted according to the locale it is addressing, which will naturally often result in slightly different messages in the same language. Practical, mostly financial, implications suggest that this cannot be easily achieved. However, some official tourism websites have started recognising the importance of locale over language. For example, at the time of writing, the official tourism website for Australia (australia.com) is available in a number of locales where English is spoken, such as USA (English), UK (English), and International (English), although an Australian version is conspicuously absent. Destination promoters, in this case, official tourism boards, are becoming more and more aware of the key role that country of origin plays in TDI formation, even though their attempts are limited to the macro-structure of the webpages and the adaptation of prominent practical information. Any adaptation of the actual textual message is, sadly, entirely missing, which suggests that micro-level considerations have not been taken into account during the localisation process and local messages are not easy to separate from overseas messages. Responsibility for this should at least partly be assigned to tourism and translation studies scholars, as there is surprisingly very little research focusing on an in-depth examination of the different communication strategies that are employed in destination promotion, how these vary across cultures, and how they can be negotiated in translation. Although a lot of attention is paid to building powerful destination brands (Morgan and Pritchard 2002), which is argued to be a long-term effort, no reference is made to the role of translation in this process. The only fleeting reference to the potentially significant role that translation might play in destination branding is found in Hallet and Kaplan-Weinger (2010), who however encourage others to explore the role of language in TDI formation, instead of doing so themselves. The starting point for this exploration is arguably the tourism text and its properties, which is what the next and final section of this chapter will focus on.

2.3 The Tourism Text

2.3.1 Functions of Tourism Texts

Tourism texts are crucial elements of the tourism process, as they mediate the relationship between tourist and destination and between host and guest (Bhattacharya 1997). Reading a tourism text prior to travelling is part of the preparation for travelling and it helps shape that all-important TDI. According to Kelly, who is a translation scholar, a tourism text is defined as "any text published by a public or private organisation of any kind intended (a) to give information to any kind of visitor or (b) to advertise a destination (city, hotel, restaurant, etc.) and encourage visitors to go there" (1997, 35). Therefore, in this account tourism texts can fulfil either an informative or a promotional function. For example, the informative function dominates in guidebooks, while leaflets and brochures have primarily a promotional function (Calvi 2010). Tourism texts might also fulfil both functions at the same time, albeit to different degrees. Seen from this perspective, official tourism websites are considered to be info-promotional, combining information and promotion, according to Valdeón (2009), who examined the translation of these texts. The main aim of such websites is "to present a location to the addressee, to deepen their understanding of it, to catch their attention and to try to persuade them to visit the location or to return there" (Klimova 2016, 169). As these websites encourage tourism, they also develop and promote for their own communities an identity "as a welcoming, soothing (divinely) poignant setting for spiritual, intellectual and cultural fulfilment" (Hallet and Kaplan-Weinger 2010, 7). The representations that inhabit tourism websites are given meaning by potential tourists, while these potential tourists construct their own identity as tourists and the identity of the locale as tourism site. This complex communicative intent of official tourism websites—as well as the fact that they have a strong impact on TDI formation, and are often translated into multiple languages—is the reason they have been chosen as a focus of study in this book. Additionally, tourism boards, which are typically responsible for the creation of official tourism websites, have been found to play a significant role in bringing suppliers and consumers closer through active promotion, not least through electronic channels (Palmer, 2002). Therefore, any references to tourism texts in this section, and the rest of book, will refer to official tourism websites.

The promotional function of tourism texts has been extensively discussed in the otherwise limited literature on the language of tourism, and persuasion is typically identified as the main function of tourism discourse (Dann 1996; Gotti 2006; Maci 2007), although this may vary according to the particular genre, as pointed out above. Tourism texts such as the official tourism websites at the centre of the current case studies aim to control attitudes and behaviours by promoting a specific TDI, therefore actively encouraging readers to become visitors. Considering that reader-tourists cannot experience the product before buying it, the tourism text aims to recreate the illusion of visiting a destination, helping them minimise the risk associated with destination selection. Conversely, the informative function of tourism texts is often neglected in existing studies of tourism discourse (although not in studies examining the translation of tourism texts as discussed in Chapter 3), which might be explained by the fact that this function is often seen as secondary: tourism texts do not aim to simply inform, but to sell, and factual information is used to enhance the attracting power of the destination. For instance, tourism texts might offer suggestions of restaurants, and examples of local dishes served. Mentioning dishes might appear as an obvious informative element (the text informs the reader-tourists about what they can eat), however there is also a strong promotional function (the text produces in the reader-tourists the desire to have a unique tasting experience). Similarly, when a city is described as *big* or a beach as *sandy*, although at first glance these descriptions might appear to be serving an informative aim, the ultimate goal is to convince the reader-tourist to visit these places. This intricate interaction between promotion and information is what makes the language of tourism particularly complex and there is little point in trying to draw a distinction between the two aims in tourism texts. For this reason, the present book focuses on info-promotion and considers the informational elements of tourism texts as supportive of their promotional function. Importantly, although the tourism text attempts to create an image of the destination through both information and promotion, what matters is how this image is interpreted by potential tourists, which relies on both contextualisation and evaluation (Bhattacharya 1997), and is dependent on their cultural background, among other things.

2.3.2 The Language of Tourism

Despite the rapid growth of the tourism industry in recent years, the heavy investment in destination promotion, and the key role played by the destination image, relatively little attention has been paid to the use of language in the tourism industry, with most studies focusing on visual material and how these represent and affect the performance of destinations (Jaworski and Pritchard 2005; Pink 2007; Govers et al. 2007;

Scarles 2010). To some extent, this lack of attention to linguistic means can be explained by the very nature of the tourism activity (Thurlow and Jaworski 2011), since it largely relies on the visual sense (Cronin 2000; Dunn 2005, 2006). However, there is strong evidence that language also plays a key role in tourism (Djafarova and Andersen 2008; Djafarova and Waring 2012). According to Thurlow and Jaworski "language [...] is everywhere in tourism; in fact, language and languages sit at the very heart of the tourist experience, its representation and its realization, its enculturation and its enactment" (2011, 289). Language in tourism is not simply a means of advertising, although language is needed in order to make tourism promotion work. According to Dann (1996, 21) "phrase precedes gaze"; it is the discourse of tourism that converts a place into a destination, that creates the idea of sight. In other words, it is because a certain place is promoted through language as having certain qualities that make it worth visiting that we start perceiving it as a touristic destination. Furthermore, the language of tourism presents a certain worldview (Knowles 1989) and allows us to locate ourselves in it (Franklin 2003). Seen that way, the language of tourism contributes to identity construction for both the destination and the potential visitor (Hallet and Kaplan-Weinger 2010). It is also crucial to the development of the idea of the tourist "following a prescribed route through a landscape of selected and ready-interpreted sites and monuments" (Gilbert 1999, 282). Studying the language of tourism can offer insights into how identities are constructed and communicated, but also into how the ordinary is transformed into the extraordinary, and how otherness becomes part of the consumer culture (Manca 2016; Hummon 1988; Francesconi 2007). Notably, as this book aims to show, the language of tourism is not a homogenous discourse across languages, which, in turn, suggests that differences are likely to, and should, be observed in the communicative means employed to negotiate identity construction.

Research in this field (Calvi 2010; Edwards and Curado 2003; Fusari 2009; Gotti 2006; Nigro 2006) focuses particularly on whether the language of tourism can be rightfully considered as a type of specialised discourse, similar to medical and legal language. Opinions differ with some scholars, like Dann (1996) and Cappelli (2007), arguing that tourism has a discourse of its own, while others, like Agorni (2012) and

Gotti (2006), question the existence of a specialised tourism discourse, claiming that it is difficult to predict the linguistic features that might be used in a tourism text, in the same way that we might expect certain linguistic features to appear in a medical or legal text. Further to this, the lexis used in tourism texts tends to lack some of the typical features of specialised vocabulary, such as monoreferentiality, lack of connotational meaning, and transparency (Gotti 2006). However, it must be noted that these features are normally associated with referential texts, while tourism texts are primarily persuasive. It has been argued that the language of tourism has a very distinct aim, similar to the language of advertising, that is to "persuade, lure, woo and seduce millions of human beings, and, in doing so, covert them from potential into actual clients" (Dann 1996, 2), and the linguistic features used to achieve this are naturally very different to the those employed in scientific, technical, or other specialised texts.

Another argument that is often put forward against the existence of a specialised tourism discourse is that it is accessible to the general public. However, according to Cappelli (2012) the seemingly everyday language used in tourism texts, although accessible, functions in a way that is similar to the language used in interactions between specialists and non-specialists. Indeed, in tourism texts, the often anonymous author is presented as an expert in a destination, "an 'observe-it-all/know-it-all' professional" (Pierini 2009, 98), addressing an audience that knows little or nothing about it. In the present book, tourism language is understood as specialised, in the sense that it relies on a number of specific linguistic features to achieve its main aim. These linguistic features, although in their majority not necessarily distinct from everyday language, are often used with a different frequency (e.g. more frequently found in tourism texts) and serve the specific aim of turning readers into visitors. At the same time, generalisations that try to identify linguistic features characterising the entire tourism discourse are avoided; the focus is rather on different elements (e.g. tourist, space) and how these are realised within the wider framework of a tourism text, following the idea put forward by Nigro (2005) and Dann (1996) that the language of tourism cannot be treated as an undifferentiated whole.

Since promotion is often considered the primary function of tourism texts, the language of tourism tends to speak only in positive and glowing terms of the attractions and services it aims to promote, and is often characterised as being very emphatic and strongly evaluative, loaded with hyperbolic language (Ip 2008). Evaluation is mainly expressed through positive adjectives (Hassan et al. 2008; Ip 2008; Maci 2007; Pierini 2009), such as *beautiful, ideal*, and *perfect*, which is the linguistic feature most typically associated with tourism discourse, together with superlatives (Ip 2008; Pierini 2009), such as *best, finest,* and *largest*. This overreliance on adjectives with strong positive connotations, at least in English, makes tourism discourse appear intentionally emphatic (Zuliani 2013) with a tendency to exaggerate.

Tourist texts also show a preference for specific verbal strategies. For instance, imperatives and directives are often used to address the reader-tourist (Ip 2008; Maci 2007; Zuliani 2013), as an invitation to enjoy what is on offer (Nigro 2006). Experiential verbs (e.g. *enjoy, see*) tend to be preferred (Maci 2007), which further strengthens the idea that the text is to be read as an open invitation to visit the destination. This element of choice is also expressed, at least in English, through the modal *can* (Maci 2007). Alternatively, the future tense is preferred, which allows reader-tourists to anticipate what will be experienced (Zuliani 2013). Focusing on the subject of these verbal structures, the second-person singular pronoun (i.e. *you*) is the most widespread option, which allows the often anonymous author of the tourism text to build a direct relationship with the audience and generate identification and loyalty (Maci 2007).

Like advertisements, tourism texts rely on alliteration, onomatopoeia, the use of familiar expressions in an unusual context, humour, and testimonies, but also on the use of keywords (such as *escape, adventure*, and *exotic*) (Dann 1996). Regarding the latter, it is important to note that these have been found to correspond to the expectations of the potential visitor, rather than the attributes of the destination (Hanefors and Larsson 1993), which is evidence that tourism texts need to be adapted according to the audience they are addressing (e.g. domestic or international). Finally, figurative language, mainly metaphors and similes, is also prominent in tourism texts, which helps communicate the intangible features of tourism (Buckley 2010; Hallet and Kaplan-Weinger 2010) and reduce unfamiliarity (Djafarova and Andersen 2008). At the same time, figurative language attracts attention and supports the creation of a positive destination image (Djafarova and Andersen 2008; Dann 1996), by reinforcing the visual image.

Certain linguistic features might also be associated with the perspective adopted in the tourism text, and the values promoted in it. In the first comprehensive examination of the language of tourism, Dann (1996) aptly summarises the theoretical approaches which have significantly contributed towards the understanding of modern tourism and its sociolinguistic correlates, and identifies four such perspectives: authenticity (e.g. traditions and local culture), strangehood (e.g. novelty, away from reality), play (e.g. hedonism and fun), and conflict (e.g. contrast between societies, often applied to tourism in developing countries). Each of these perspectives highlights different aspects of the tourism experience expressed through different choices of lexis. For example, authenticity might be conveyed by adjectives such as *actual, authentic,* and *real*, while strangehood though adjectives like *different, unique,* and *remote*. Similarly, some scholars, like Maci (2007), identify certain values that characterise destinations, such as preservation (e.g. *natural, untouched*), continuity (e.g. *traditional, timeless*), novelty (e.g. *exotic, adventurous*), distance (e.g. *remote, isolated*), exclusiveness (e.g. *exclusive, unique*), and attractiveness (e.g. *picturesque, pretty*). Studying the lexis employed in tourism texts can therefore reveal not only the strategy adopted for tourism promotion, but also how a destination identity is selected and communicated. For instance, a study conducted by White (2007) reveals that Fiji tourism discourse focuses on authenticity and otherness by highlighting the 'primitiveness' and 'friendliness' of Fijians as indigenous people, features which are believed to be attractive to Western tourists, through phrases like *genuine friendliness of the Fijian people, the friendliest people on earth,* and *Fijians were born with an innate sense of hospitality*. Dann (1996) also identifies four themes typically employed in tourism texts which are: (a) romanticism, regression, and rebirth; (b) happiness, hedonism, and heliocentrism; (c) fun, fantasy, and fairy tales; and (d) sea, sex, and socialisation. Each text is likely to capitalise on one of these themes, although some themes might be combined, for instance, the second and the fourth one. What all these themes have in common, according to Dann, is that they address the potential tourist as a child so that they can be more easily controlled.

The majority of studies on the language of tourism examined here focus on English tourism texts, and therefore, any generalisation can only apply to texts produced in that language, and even so, cultural differences need to be taken into account relating to the specific variety of English (e.g. British or American). However, the investigation of the language of advertising and the discussion of the personal factors affecting TDI suggest that the communicative strategies adopted by destination promoters need to be adapted depending on the audience and its cultural background. Different languages will adopt different strategies in promoting a destination depending on their unique features and how these are perceived (Hassan et al. 2008). Examining the textual approaches employed across languages allows for a much deeper understanding of the nature of tourism (Djafarova and Andersen 2008) and how exactly promotion is achieved in it. Although there is some implicit recognition that the language of tourism needs to address potential tourists "in terms of their own *culturally predicated* needs and motivations" (Dann 1996, 2 emphasis added), there is generally very little focus on the cross-linguistic comparison of tourism texts. The next chapter explores previous research focusing on the cross-linguistic differences in the language of tourism, focusing specifically on the key role that translation plays.

References

Abernethy, Avery, and George Franke. 1996. "The Information Content of Advertising: A Meta- Analysis." *Journal of Advertising* 25 (2): 1–17.

Agapito, D., P. Valle, and J. Mendes. 2013. "The Cognitive-Affective-Conative Model of Destination Image: A Confirmatory Analysis." *Journal of Travel & Tourism Marketing* 30 (5): 471–81.

Agorni, Mirella. 2012. "Tourism Communication: The Translator's Responsibility in the Translation of Cultural Difference." *PASOS: Revista de Turismo y Patrominio Cultural* 10 (4): 5–11.

Ahn, Jungsun, Carrie La Ferle, and Doohwang Lee. 2016. "Language and Advertising Effectiveness: Code-Switching in the Korean Marketplace." *International Journal of Advertising* 36 (3): 477–95.

Alpert, Mark, Judy Alpert, and Elliot Maltz. 2005. "Purchase Occasion Influence on the Role of Music in Advertising." *Journal of Business Research* 58: 369–76.

Anderson, C. 2006. *The Long Tail: Why the Future of Business Is Seling Less of More*. New York: Hyperion Books.

Andreu, Luisa, J. Enrique Bigné, and Chris Cooper. 2000. "Projected and Perceived Image of Spain as a Tourist Destination for British Travellers." *Journal of Travel & Tourism Marketing* 9 (4): 47–67.

Bagozzi, R. 1975. "Marketing as Exchange." *Journal of Marketing* 39 (4): 32 39.

Baloglu, S., and D. Brinberg. 1997. "Affective Images of Tourism Destinations." *Journal of Travel Research* 35 (4): 11–15.

Baloglu, S., and K. McCleary. 1999. "A Model of Destination Image Formation." *Annals of Tourism Research* 26: 868–97.

Bastida, U., and T. Huan. 2014. "Performance Evaluation of Tourism Websites' Information Quality of Four Global Destination Brands: Beijing, Hong Kong, Shanghai, and Taipei." *Journal of Business Research* 67 (2): 167–70.

Baumgardner, R. 2006. "The Appeal of English in Mexican Commerce." *World Englishes* 25: 251– 66.

Beerli, A., and J. Martín. 2004. "Factors Influencing Destination Image." *Annals of Tourism Research* 31 (3): 657–81.

Belch, G., and M. Belch. 2001. *Advertising and Promotion: An Integrated Marketing Communications Perspective*. New York: McGraw-Hill.

Bennett, P. D. 1995. *Dictionary of Marketing Terms*. Edited by NTC Business Books. Chicago.

Berlyne, Daniel. 1971. *Aesthetics and Psychobiology*. New York: Appleton.

Bhattacharya, D. 1997. "Mediating India: An Analysis of a Guidebook." *Annalsa of Tourism Research* 24: 371–89.

Bigné, J. E., and M. Zorio. 1989. "Marketing Turístico: El Proceso de Toma de Decisiones Vacacionales." *Revista de Economia y Empresa* 9 (23): 91–112.

Bigné, J. E., M. I. Sanchez, and J. Sanchez. 2001. "Tourist Image, Evaluation Variables and After Purchase Behaviour: Inter-Relationship." *Tourism Management* 22: 607–16.

Bishop, M., and M. Peterson. 2010. "The Impact of Medium Context on Bilingual Consumers' Responses to Code-Switched Advertising." *Journal of Advertising* 39 (3): 55–67.

Bolen, William. 1984. *Advertising*. 2nd ed. New York: Wiley.

Bosomworth, D. (2012, October 17). *How to Give Your Content Wings*. Smart Insights. https://www.smartinsights.com/content-management/content-marketing-strategy/how-to-give-your-content-wings/. Accessed 12 March 2023.

Boulding, K. 1956. *The Image-Knowledge in Life and Society*. Ann Arbor, MI: University of Michigan Press.

Brierley, Sean. 1995. *The Advertising Handbook*. Oxon and New York: Routledge.

Brokaw, S. G. 1990. "An Investigation of Jewelry Store Image Structure." Florida State University, Tallahassee.

Bruthiaux, Paul. 2000. "In a Nutshell: Persuasion in the Spatially Constrained Language of Advertising." *Language & Communication* 20: 297–310.

Buckley, R. 2010. "Communications in Adventure Tour Products: Health and Safety in Rafting and Kayaking." *Annals of Tourism Research* 37 (2): 315–32.

Buhalis, D., and R. Law. 2008. "Progress in Information Technology and Tourism Management: 20 Years on and 10 Years After Internet—The State of ETourism Research." *Tourism Management* 29 (4): 609–23.

Bulmer, Sandy, and Margo Buchanan-Oliver. 2006. "Visual Rhetoric and Global Advertising Imagery." *Journal of Marketing Communications* 12 (1): 49–61.

Burdick, Anne, Johanna Drucker, Peter Lunenfeld, Todd Presner, and Jeffrey Schnapp. 2012. *Digital Humanities*. Cambridge, MA: MIT Press.

Cai, L., B. Wu, and B. Bai. 2004. "Destination Image and Loyalty." *Tourism Review International* 7 (3–4): 153–62.

Callow, M., and L. Schiffman. 1999. "A Visual Esperanto? The Pictorial Metaphor in Global Advertising." *European Advances in Consumer Research* 4: 17–20.

Calvi, M. V. 2010. "Los Generos Discursivos En La Lengua Del Turismo: Una Propuesta de Clasificación." *Ibérica* 19: 9–32.

Cappelli, Gloria. 2007. *Sun, Sea, Sex and the Unspoilt Countryside*. 2nd ed. Pari: Pari Publishing.

———. 2012. "Travelling in Space: Spatial Representation in English and Italian Tourism Discourse." *Textus: English Studies in Italy* 1: 19–36.

Caroll, R., and David Luna. 2011. "The Other Meaning of Fluency: Content Accessibility and Language in Advertising to Bilinguals." *Journal of Advertising* 40 (3): 73–94.

Chen, P., and D. Kerstetter. 1999. "International Students' Image of Rural Pennsylvania as a Travel Destination." *Journal of Travel Research* 37: 256–66.

Chen, C., and D. Tsai. 2007. "How Destination Image and Evaluative Factors Affect Behavioural Intentions." *Tourism Management* 28: 1115–22.

Chernatony, L. De, and M. McDonald. 1992. *Creating Powerful Brands: The Strategic Route to Success in Consumer, Industrial and Service Markets*. Oxford: Butterworth.

Chiou, Jyh-shen. 2002. "The Effectiveness of Different Advertising Message Appeals in the Eastern Emerging Society: Using Taiwanese TV Commercials as an Example." *International Journal of Advertising* 21 (2): 217–36.

Choi, S., X. Y. Lehto, and A. M. Morrison. 2007. "Destination Image Representations on the Web: Content Analysis of Macau Travel Related Websites." *Tourism Management* 28: 118–29.

Chon, K. S. 1990. "The Role of Destination Image in Tourism: A Review and Discussion." *The Tourist Review* 45 (2): 2–9.

Chowdhury, Rafi, Douglas Olsen, and John Pracejus. 2011. "How Many Pictures Should Your Print Ad Have?" *Journal of Business Research* 34: 3–6.

Christodoulides, G., and C. Jevons. 2011. "Teh Voice of the Consumer Speaks Forcefully in Brand Identity: User-Generated Content Forces Smart Marketers to Listen." *Journal of Advertising Research* 51 (1): 101–11.

Christodoulides, G., C. Jevons, and J. Bonhomme. 2012. "Memo to Marketers: Quantitative Evidence for Change: How User-Generated Content Really Affects Brands?" *Journal of Advertising Research* 52 (1): 53–64.

Cook, Guy. 2001. *The Discourse of Advertising*. 2nd ed. Oxon and New York: Routledge.

Corbett, Edward. 1990. *Classical Rhetoric for the Modern Student*. New York: Oxford University Press.

Crompton, J. L. 1979. "An Assessment of the Image of the Mexico as a Vacation Destination and the Influence of Geographical Location upon the Image." *Journal of Travel Research* 17 (4): 18–23.

Crompton, P., and R. McAlea. 2000. "Rhetorical Devices in Television Advertising." In *Advertising and Identity in Europe: The I of the Beholder*, edited by J. Cannon, P. Odber de Baubeta, and R. Warner, 32–41. Bristol, UK: Intellect Books.

Cronin, Michael. 2000. *Across the Lines: Travel, Language, Translation*. Cork: Cork University Press.

Cutler, Bob, Rajshekhar Javalgi, and Krishna Erramilli. 1992. "The Visual Component of Print Advertising: A Five-Country Cross-Cultural Analysis." *European Journal of Marketing* 26 (4): 7–20.

Dann, Graham M. S. 1996. *The Language of Tourism*. Oxon: CAB International.

Djafarova, Elmira, and H. Andersen. 2008. "Contribution of Figurative Devices to Representation of Tourism Images." *Journal of Vacation Marketing* 14 (4): 291–303.

Djafarova, Elmira, and Teresa Waring. 2012. "Language of Tourism Advertising: A Pragmatic Approach." *Tourism Analysis* 17: 233–37.

Domzal, T., J. Hunt, and J. Kernen. 1995. "Achtung! The Information Processing of Foreign Words in Advertising." *International Journal of Advertising* 14: 94–114.

Dragova, S., K. Petrovskaya, and R. Egger. 2014. "Measuring the Perceived Image of Lithuania through Its Destination Management Organization Website." In *Information and Communication Technologies in Tourism 2014*, edited by Z. Xiang and I. Tussyadiah, 679–91. Cham: Springer.

Dorfman, Peter, and John Howell. 1988. "Dimension of National Culture and Effective Leadership Patterns: Hofstede revisited". In *Advanced in International Comparative Management: A research annual*, edited by R. Farmer and E. Goun, 127–49. Stamford: JAI Press.

Dunn, David. 2005. "'We Are Not Here to Make a Film About Italy, We Are Here to Make A Film About Me...' British Television Holiday Programmes' Representations of the Tourist Destination." In *The Media and Tourist Imagination: Converging Cultures*, edited by David Crouch, Rhona Jackson, and Felix Thompson, 154–69. London.

———. 2006. "Singular Encounters: Mediating the Tourist Destination in British Television Holiday Programmes." *Tourist Studies* 6: 37–58.

Edwards, Patricia, and Alejandro Curado. 2003. "The Promotion of Tourism Through Key Concepts and Specific Discourse." *LSP & Professional Communication* 3 (1): 26–42.

Eurobarometer. 2021. "Flash Eurobarometer 499: Attitudes of Europeans Towards Tourism." Brussels.

Fernández Cavia, J., C. Rovira, P. Díaz-Luque, and C. Cavaller. 2014. "Web Quality Index (WQI) for Official Tourist Destination Websites: Proposals for an Assessment System." *Tourism Management Perspectives* 9: 5–13.

Fesenmaier, D., and S. Cook. 2009. *Travellers' Use of the Internet*. Washington: U.S. Travel Association.

Francesconi, Sabrina. 2007. *English for Tourism Promotion: Italy in British Tourism Texts*. Milano: Hoepli.

———. 2014. *Reading Tourism Texts: A Multimodal Analysis*. Briston, Tonawanda, and Ontario: Channel View Publications.

Franklin, A. 2003. *Tourism: An Introduction*. London: SAGE Publications.

Frias, D. M., M. A. Rodríguez, J. Alberto Castañeda, C. M. Sabiote, and D. Buhalis. 2011. "The Formation of a Tourist Destination's Image via Information Sources: The Moderating Effect of Culture." *International Journal of Tourism Research* 14 (5): 437–50.

Fuertes-Olivera, Pedro, Marisol Velasco-Sacristan, Ascensión Arribas-Baño, and Eva Samaniego-Fernández. 2001. "Persuasion and Advertising English: Metadiscourse in Slogans and Headlines." *Journal of Pragmatics* 33 (8): 1291–1307.

Fusari, Sabrina. 2009. "Multilingual Tourist Videos as Specialized Discourse: The Case Study of Val Gardena." In *Quaderni Del CeSLiC, Occasional Papers*, edited by Donna R Miller, 4–26. Bologna: Centro di Studi Linguistico-Culturali.

Galí, N., and J. A. Donnaire. 2005. "The Social Construction of the Image of Girona: A Methodological Approach." *Tourism Management* 26 (5): 777–85.

Gallarza, M., G. Irene, and H. Calderón. 2002. "Destination Image Towards a Conceptual Framework." *Annals of Tourism Research* 29 (1): 56–78.

Gardner, Rod, and Sigrid Luchtenberg. 2000. "Reference, Image, Text in German and Australian Advertising Posters." *Journal of Pragmatics* 32: 1807–21.

Gartner, W. 1989. "Tourism Image: Attribute Measurement of State Tourism Products Using Multidimensional Scaling Techniques." *Journal of Travel Research* 28 (2): 16–20.

———. 1993. "Image Formation Process." *Journal of Travel & Tourism Marketing* 2 (2–3): 191–216.

Gilbert, D. 1999. "'London in All Its Glory—Or How to Enjoy London' Guidebok Representations of Imperial London." *Journal of Historical Geography* 25 (3): 279–97.

Gitelson, R., and D. Kerstetter. 1994. "The Influence of Friends and Relatives in Travel Decision-Making." *Journal of Travel & Tourism Marketing* 3 (3): 59–68.

Gotti, Maurizio. 2006. "The Language of Tourism as Specialized Discourse." In *Translating Tourism: Linguistic/Cultural Representations*, edited by Oriana Palusci and Sabrina Francesconi. Trento: Trento University Press.

Govers, R., and F. M. Go. 2005. "Projected Destination Image Online: Website Content Analysis of Pictures and Text." *Information Technology and Tourism* 7: 73–89.

Govers, Robert, Frank M. Go, and Kuldeep Kumar. 2007. "Promoting Tourism Destination Image." *Journal of Travel Research* 46: 15–23.

Grosjean, F. 1982. *Life with Two Languages: An Introduction to Bilingualism.* Cambridge: Harvard University Press.
Gunn, C. 1972. *Vacationscape: Designing Tourist Regions.* Austin: Bureau of Business Research, University of Texas.
Gursoy, D., and J. Chen. 2000. "Competitive Analysis of Cross-Cultural Information Search Behaviour." *Tourism Management* 21 (6): 583–90.
Gursoy, D., and K. McCleary. 2004. "An Integrative Model of Tourists' Information Search Behavior." *Annals of Tourism Research* 31: 353–73.
Gursoy, D., and W. T. Umbreit. 2004. "Tourist Information Search Behaviour: Cross-Cultural Comparison of European Union Member States." *The International Journal of Hospitality Management* 23: 55–70.
Guthrie, J., and P. Gale. 1991. "Positioning Ski Areas." In *New Horizons Conference Proceedings*, 551–69. Calgary: University of Calgary.
Hallet, R. W., and J. Kaplan-Weinger. 2010. *Official Tourism Websites: A Discourse Analysis Perspective.* Bristol: Channel View Publications.
Hanefors, M., and L. Larsson. 1993. "Video Strategies Used by Tour Operators: What Is Really Communicated?" *Tourism Management* 14 (1): 27–33.
Hassan, Hanita, Hadina Habil, and Zaliza Mohd Nasir. 2008. "Persuasive Strategies of Tourism Discourse." In *New Perspectives in Language and Communication Research*, edited by Hadina Habil and Hanita Hassan, 1–20. Skudai, Malaysia: Universiti Teknologi Malaysia.
Hofstede, G. 2001. *Culture's Consequences: International Differences in Work Related Values.* 2nd ed. Newbury Park, California: SAGE Publications.
Hornikx, Jos, Frank Van Meurs, and Robert-jan Hof. 2013. "The Effectiveness of Foreign-Language Display in Advertising for Congruent Versus Incongruent Products." *Journal of International Consumer Marketing* 25 (3): 152–65.
Hsieh, A. T., and C.-W. Tsai. 2009. "Does Culture Really Matter? Hotel Service Perceptions by Taiwan and American Tourists." *Journal of Culture, Tourism and Hospitality* 3 (1): 54–69.
Huang, C. Y., C. J. Chou, and P. C. Lin. 2010. "Involvement Theory in Constructing Bloggers' Intention to Purchase Travel Products." *Tourism Management* 32: 513–26.
Hummon, D. 1988. "Tourist Worlds: Tourist Advertising, Ritual, and American Culture." *The Sociological Quarterly* 29 (2): 149–337.
Hung, K. 2000. "Narrative Music in Congruent and Incongruent TV Advertising." *Journal of Advertising* 29: 26–34.
———. 2001. "Framing Meaning Perceptions with Music: The Case of Teaser Ads." *Journal of Advertising* 30: 39–49.

Hunt, J. 1975. "Image as a Factor in Tourism Development." *Journal of Travel Research* 13 (3): 1–17.

Hwang, Y., U. Gretzel, Z. Xiang, and D. Fesenmaier. 2006. "Information Search for Travel Decisions." In *Destination Recommendation Systems: Behavioral Foundations and Applications*, edited by R. Fesenmaier, H. Werthner, and K. Wöber, 3–16. Cambridge: CAB International.

Ip, J. Y. L. 2008. "Analyzing Tourism Discourse: A Case of a Hong Kong Travel Brochure." *LCOM Papers* 1: 1–19.

Jacobsen, J., and A. Munar. 2012. "Tourist Information Search and Destination Choice in a Digital Age." *Tourism Management Perspectives* 1: 39–47.

Jaworski, A., and A. Pritchard. 2005. *Discourse, Communication and Tourism.* Clevedon: Channel View Publications.

Jiménez-Crespo, M. Á. 2013. *Translation and Web Localization.* London: Routledge.

Kelly, Dorothy. 1997. "The Translation of Texts from the Tourist Sector: Textual Conventions, Cultural Distance and Other Constrains." *TRANS* 2: 33–42.

Kelly-Holmes, H. 2000. "Bier, Parfum, Kaas: Language Fetish in European Advertising." *European Journal of Cultural Studies* 3: 67–82.

———. 2005. *Advertising as Multilingual Communication.* New York: Palgrave Macmillan.

Khan, A. 2014. "Social Aspects of Code-Switching: An Analysis of Pakistani Television Advertisements." *Information Management and Business Review* 6: 269–79.

Kim, H., and D. Fesenmaier. 2008. "Persuasive Design of Destination Websites: An Analysis of First Impressions." *Journal of Travel Research* 47: 3–13.

Kim, S., and X. Y. Lehto. 2013. "Projected and Perceived Destination Brand Personalities: The Case of South Korea." *Journal of Travel Research* 52 (1): 117–30.

Kim, S., and Y. Yoon. 2003. "The Hierarchical Effects of Affective and Cognitive Components on Tourism Destination Image." *Journal of Travel & Tourism Marketing* 14 (2): 1–22.

Kim, J., J.-S. Lim, and M. Bhargava. 1998. "The Role of Affect in Attitude Formation: A Classical Conditioning Approach." *Journal of the Academy of Marketing Science* 26 (2): 143–52.

Klimova, Katarina. 2016. "Italy: Web Tourism Promotion." *Italiano LinguaDue* 2: 168–80.

Knowles, Murray. 1989. "Some Characteristics of a Special Language: The Language of Tourism." In *From Ofice to School: Special Language & Internationlisation*, edited by Christer Laurén and Marianne Nordman, 59–66. Clevedon: Multilingual Matters.
Konecnik, M., and W. Gartner. 2007. "Customer-Based Brand Equity for a Destination." *Annals of Tourism Research* 34 (2): 400–21.
Kotler, Philip, and S. Levy. 1969. "Broadening the Concept of Marketing." *Journal of Marketing* 33: 10–15.
Kozak, M. 2002. "Comparative Analysis of Tourist Motivations by Nationality and Destinations." *Tourism Management* 23: 221–32.
Kroeber-Riel, Werner. 1992. "Globalisierung Der Euro- Werbung. Ein Konzeptioneller Ansatz Der Konsumentenforschung." *Marketing ZFFP* 14 (4): 261–67.
Leclerc, F., B. Schmitt, and L. Dube. 1994. "Foreign Branding and Its Effects on Product Perceptions and Attitudes." *Journal of Marketing Research* 31: 263–70.
Leech, Geoffrey. 1966. *English in Advertising: A Linguistic Study of Advertising in Great Britain.* London: Longman.
Lepp, A., H. Gibson, and C. Lane. 2011. "Image and Perceived Risk: A Study of Uganda and Its Official Tourism Website." *Tourism Management* 32 (3): 675–84.
Lin, Carolyn. 2001. "Cultural Values Reflected in Chinese and American Television Advertising." *Journal of Advertising* 30 (4): 83–94.
Llodrà-Riera, I., M. P. Martínez-Ruiz, and A. Jiménez-Zarco, A. I. Izquierdo-Yusta. 2015. "A Multidimensional Analysis of the Information Sources Construct and Its Relevance for Destination Image Formation." *Tourism Management* 48: 319–28.
Loda, M., K. Teichmann, and A. Zins. 2009. "Destination Websites' Persuasiveness." *International Journal of Culture, Tourism and Hospitality Research* 3 (1): 70–80.
Luna, David, and Laura Peracchio. 1999. "What's in a Bilingual's Mind? How Bilingual Consumers Process Information." In *Advances in Consumer Research*, edited by Eric Arnould and Linda Scott, 306–11. Provo: Association for Consumer Research.
———. 2001. "Moderators of Language Effects in Advertising to Bilinguals: A Psycholinguistic Approach." *Journal of Consumer Research* 28 (2): 284–95.
———. 2002. "'Where There Is a Will…': Motivation as a Moderator of Language Processing by Bilingual Consumers." *Psychology & Marketing* 19 (7–8): 573–93.

———. 2005a. "Advertising to Bilingual Consumers: The Impact of Code-Switching on Persuasion." *Journal of Consumer Research* 31 (4): 760–65.

———. 2005b. "Sociolinguistic Effects on Code-Switched Ads Targeting Bilingual Consumers." *Journal of Advertising* 34 (2): 43–56.

Luna, David, D. Lerman, and Laura Peracchio. 2005. "Structural Constraints in Code-Switched Advertising." *Journal of Consumer Research* 32: 416–23.

Maci, Stefania. 2007. "Virtual Touring: The Web-Language of Tourism." *Linguistica e Filologia* 25: 41–65.

MacInnis, D. J., and C. W. Park. 1991. "The Differential Role of Characteristics of Music on High- and Low-Involvement Consumers." *Journal of Consumer Research* 18: 161–73.

MacKay, Kelly, and Daniel Fesenmaier. 2000. "An Exploration of Cross-Cultural Destination Image Assessment." *Journal of Travel Research* 38: 417–23.

Manca, Elena. 2016. "Official Tourist Websites and the Cultural Communication Grammar Model: Analysing Language, Visuals, and Cultural Features." *Cultus: The Journal of Intercultural Mediation and Communication* 9 (1): 2–22.

Mansfield, Y. 1992. "Motivation to Actual Travel." *Annals of Tourism Research* 19: 399–419.

Marine-Roig, Estela, and S. A. Clavé. 2016. "Destination Image Gaps Between Official Tourism Websites and User-Generated Content." In *Information and Communication Technologies in Tourism 2016*, edited by A. Inversini and R. Schegg, 253–65. Cham: Springer.

Martenson, Rita. 1987. "Advertising Strategies and Information Content in American and Swedish Advertising: A Comparative Content Analysis in Cross-Cultural Copy Research." *International Journal of Advertising* 6: 133–44.

Martin, Elizabeth. 2008. "Language-Mixing in French Print Advertising." *Journal of Creative Communications* 3 (1): 49–76.

Mattila, A., and J. Wirtz. 2001. "Congruency of Scent and Music as a Driver of Instore Evaluations and Behaviour." *Journal of Retailing* 77: 273–89.

Mayo, E. J. 1973. "Regional Images and Regional Travel Behaviour." In *Proceedings of the Travel Research Association 4th Annual Conference*, 211–18.

McQuarrie, Edward, and David Mick. 1996. "Figures of Rhetoric in Advertising Language." *Journal of Consumer Research* 22 (4): 424–38.

Mohammed, S. N. 2004. "Self-Representation of Small Developing Countries on the World Wide Web." *New Media and Society* 6 (4): 469–86.

Money, R. B., and J. C. Crotts. 2003. "The Effect of Uncertainty-Avoidance on Information Search, Planning, and Purchases of International Travel Vacations." *Tourism Management* 24: 191–202.

Morgan, N., and A. Pritchard. 2002. "Contextualising Destination Branding." In *Destination Branding: Creating the Unique Destination Proposition*, edited by N. Morgan, A. Pritchard, and R. Pride, 11–41. Oxford: Buterworth Heinemann.

Morgan, N., A. Pritchard, and R. Pride. 2002. "Introduction." In *Destination Branding: Creating the Unique Destination Proposition*, edited by N. Morgan, A. Pritchard, and R. Pride, 3–10. Oxford: Butterworth Heinemann.

Moriarty, Sandra, and Thomas Duncan. 1991. "Global Advertising: Issues and Practices." *Current Issues and Research in Advertising* 13 (2): 313–41.

Mulken, M. van. 2003. "Analyzing Rhetorical Devices in Print Advertisements." *Document Design* 4 (2): 114–28.

Murray, Noel, and Sandra Murray. 1996. "Music and Lyrics in Commercials: A Cross-Cultural Comparison Between Commercials Run in the Dominican Republic and in the United States." *Journal of Advertising* 25 (2): 51–63.

Myers, Greg. 1994. *Words in Ads*. London: Arnold.

Myers-Scotton, C. 1993. *Social Motivations for Code Switching: Evidence from Africa*. Oxford and New York: Oxford University Press.

Nadeau, J., L. Heslop, N. O'Reilly, and P. Luk. 2008. "Destination in a Country Image Context." *Annals of Tourism Research* 35 (1): 84–106.

Newman, Karen, and Stanley Nollen. 1996. "Culture and Congruence: The Fit Between Management Practices and National Culture." *Journal of International Business Studies* 27(4): 753–79.

Nigro, Maria Giovanna. 2005. "The Language of Tourism as LSP? A Corpus-Based Study of the Discourse of Guidebooks." In *Modern Approaches to Terminological Theories and Applications*, edited by H. Picht, 187–97. Bern and Oxford: Peter Lang.

———. 2006. *Il Linguaggio Specialistico Del Turismo: Aspetti Storici, Teorici e Traduttivi*. Roma: Aracne.

Noriega, Jaime, and Edward Blair. 2008. "Advertising to Bilinguals: Does the Language of Advertising Influence the Nature of Thoughts?" *Journal of Marketing* 72: 69–83.

Oakes, Steve. 2007. "Evaluating Empirical Research into Music in Advertising: A Congruity Perspective." *Journal of Advertising Research* 47 (1): 38–50.

Palmer, Adrian. 2002. "Destination Branding and the Web." In *Destination Branding: Creating the Unique Destination Proposition*, edited by N. Morgan, A. Pritchard, and R. Pride, 186–98. Oxford: Butterworth Heinemann.

Petty, R., and J. Cacioppo. 1991. "Effects of Issue Involvement on Attitudes in an Advertising Context." In *Proceedings of the Division 23 Program*, edited by G. Gom and M. Goldberg, 75–79. Montreal: American Psychological Association.

Phau, Ian, Tekle Shanka, and Neema Dhayan. 2010. "Destination Image and Choice Intention of University Student Travellers to Mauritius." *International Journal of Contemporary Hospitality Management* 22 (5): 758–64.

Phillips, Barbara, and Edward McQuarrie. 2004. "Beyond Visual Metaphor: A New Typology of Visual Rhetoric in Advertising." *Marketing Theory* 4: 111–34.

Pierini, Patrzia. 2009. "Adjectives in Tourism English on the Web: A Corpus-Based Study." *CÍRCULO de Lingüística Aplicada a La Comunicación (Clac)* 40: 93–116.

Pieters, R., and M. Wedel. 2004. "Attention Capture and Transfer in Advertising: Brand, Pictorial and Text Size Effects." *Journal of Marketing* 68 (2): 36–50.

Pike, S., and C. Ryan. 2004. "Destination Positioning Analysis Through a Comparison of Cognitive, Affective, and Conative Perceptions." *Journal of Travel Research* 42 (4): 333–42.

Pink, S. 2007. *The Future of Visual Anthropology: Engaging the Senses*. London: Routledge.

Pritchard, Annette, and Nigel Morgan. 1998. "'Mood Marketing'—The New Destination Branding Strategy: A Case Study of 'Wales' The Brand." *Journal of Vacation Marketing* 4: 215–29.

Ray, N., M. Ryder, and S. Scott. 1991. "Toward an Understanding of the Use of Foreign Words in Print Advertising." *Journal of International Consumer Marketing* 3: 69–97.

Reisinger, Y., and L. Turner. 1998. "Cultural Differences Between Mandarin-Speaking Tourists and Australian Hosts and Their Impact on Cross-Cultural Tourist-Host Interaction." *Journal of Business Research* 42: 175–87.

Rodríguez-Molina, M. A., D. M. Frías-Jamilena, and J. A. Castañeda-García. 2015. "The Contribution of Website Design to the Generation of Tourist Destination Image: The Moderating Effect of Involvement." *Tourism Management* 47: 303–17.

Scarles, C. 2010. "Where Words Fail Visual Ignite." *Annals of Tourism Research* 37 (4): 905–26.

Schreyer, R. D., D. Lime, and D. Williams. 1984. "Characterizing the Influence of Past Experience on Recreation Behaviour." *Journal of Leisure Research* 16: 34–50.

Schwartz, S. 1999. "A Theory of Cultural Values: Some Implications for Work." *Applied Psychology* 48 (1): 23–47.

Scott, Linda. 1990. "Understanding Jingles and Needledrop: A Rhetorical Approach to Music in Advertising." *Journal of Consumer Research* 17: 223–36.

———. 1994. "Images in Advertising: The Need for a Theory of Visual Rhetoric." *Journal of Consumer Research* 21: 252–73.

Sirgy, J., and Chenting Su. 2000. "Destination Image, Self-Congruity, and Travel Behavior: Toward an Integrative Model." *Journal of Travel Research* 38: 340–52.

Smith, S. L. 1994. "The Tourism Product." *Annals of Tourism Research* 21: 582–95.

Stern, Barbara. 1988. "Figurative Language in Services Advertising: The Nature and Uses of Imagery." In *Advances in Consumer Research*, edited by Michael Houston, 185–90. Provo: Association for Consumer Research.

Stern, Bruce, and Alan Resnik. 1991. "Information Content in Advertising: A Replication and Extension." *Journal of Advertising Research* 31 (3): 36–46.

Tasci, A., and W. Gartner. 2007. "Destination Image and Its Functional Relationships." *Journal of Travel Research* 45 (4): 413–25.

Tasci, A., W. Gartner, and S. Cavusgil. 2007. "Conceptualization and Operationalization of Destination Image." *Journal of Hospitality & Tourism Research* 31 (2): 194–223.

Taylor, Timothy. 2012. *The Sounds of Capitalism: Advertising, Music, and the Conquest of Culture*. Chicago: The University of Chicago Press.

Therkelsen, Annette. 2003. "Imagining Places: Image Formation of Tourists and Its Consequences for Destination Promotion." *Scandinavian Journal of Hospitality and Tourism* 3 (2): 134–50.

Thurlow, Crispin, and Adam Jaworski. 2011. "Tourism Discourse: Languages and Banal Globalization." *Applied Linguistics Review* 2: 285–312.

Tixier, Maud. 1992. "Comparison of the Linguistic Message in Advertisements According to the Criteria of Effective Writing." *International Journal of Advertising* 11: 139–55.

Usunier, Jean-Claude. 1996. "Atomistic Versus Organic Approaches." *International Studies of Management and Organization* 26 (4): 90–112.

———. 2006. "Relevance in Business Research: The Case of Country-of-Origin Research in Marketing." *European Management Review* 3 (1): 60–73.

Usunier, Jean-Claude, and Julie Anne Lee. 2005. *Marketing Across Cultures*. 4th ed. Harlow: Pearson.

Uysal, M., C. D. McDonald, and L. J. Reid. 1990. "Sources of Information Used by International Visitors to US Parks and Natural Areas." *Journal of Park and Recreation Administration* 8 (1): 51–59.

Valdeón, Roberto A. 2009. "Info-Promotional Material Discourse and Its Translation: The Case of the Asturias Tourist Board Texts." *Across Languages and Cultures* 10 (1): 21–47.

Vettorel, P. 2013. "English in Italian Advertising." *World Englishes* 32 (2): 261–78.

Walmsley, D. J., and J. M. Jenkins. 1993. "Appraisive Images of Tourist Areas: Application of Personal Construct." *Australian Geographer* 24 (2): 1–13.

White, C. 2004. "Destination Image: To See or Not to See." *International Journal of Contemporary Hospitality Management* 16: 309–14.

———. 2007. "More Authentic than Thou." *Tourist Studies* 7 (1): 25–49.

World Tourism Organization. 2001. *Tourism 2020*. Madrid: WTO.

Wyer, Robert. 2002. "Language and Advertising Effectiveness: Mediating Influences of Comprehension and Cognitive Elaboration." *Psychology & Marketing* 19 (7–8): 693–712.

Xue, Fei. 2014. "It Looks Green: Effects of Green Visuals in Advertising on Chinese Consumers' Brand Perception." *Journal of International Consumer Marketing* 26 (1): 75–86.

You, X., J. O. Leary, A. Morrison, and A. Hong. 2001. "A Cross Cultural Comparison of Travel Push and Pull Factors." *International Journal of Hospitality and Tourism Administration* 1 (3): 1–26.

Yu, Y., and A. Dean. 2001. "The Contribution of Emotional Satisfaction to Consumer Loyalty." *International Journal of Service Industry Management* 12 (3): 234–50.

Zander, Mark. 2006. "Musical Influences in Advertising: How Music Modifies First Impressions of Product Endorsers and Brands." *Psychology of Music* 34 (4): 465–80.

Zandpour, Fred, Cypress Chang, and Joelle Catalano. 1992. "Stories, Symbols and Straight Talk: A Comparative Analysis of French, Taiwanese and U.S. TV Commercials." *Journal of Advertising Research* 32: 25–37.

Zuliani, Alessandra. 2013. "A Journey to Portugal: The Official Tourist Promotion of Turismo de Portugal." In *Tourism and Tourist Promotion around the World*, edited by Elena Manca and F. Bianchi, 61–80. Lecce, Italy: Università del Salento.

3

Translating Tourism

3.1 Cross-Linguistic Differences in the Language of Tourism

Although it is generally accepted that tourism texts need to be adapted through translation to address a new audience, research on how this might be achieved is generally limited and patchy. There is also the assumption that, as part of the tourism experience, readers might even enjoy the different cultural features encoded in the communicative and stylistic conventions of the tourism text. In other words, they do not require translations to be adapted to the conventions they are used to in their target language. However, this assumption has been said to be both erroneous and dangerous, as, during the stage of pre-visit, readers are not yet tourists, but potential customers, who need to be convinced about the tourism product using appropriate linguistic strategies (Manca 2011). Research on the linguistic differences in the language of tourism is of particular importance for translation research, as it provides the necessary evidence that, despite some similarities, there is no universal discourse of tourism, and therefore linguistic adaptation is necessary. This is the reason why, with very few exceptions of comparative studies, such research is closely linked to translation.

S. Malamatidou, *Translating Tourism*, Palgrave Studies in Translating and Interpreting, https://doi.org/10.1007/978-3-031-49349-2_3

Before I examine previous studies focusing on the translation of tourism texts, it is necessary to clarify here what is meant by 'translation' and 'adaptation' in the present book. A broad definition of translation, which is prevalent in translation studies, encompasses adaptation and in that sense all translations are to a larger or smaller degree linguistic adaptations, that is, linguistic changes need to be made for the target text to fulfil the aim of the brief. Adaptation, on the other hand, is often used to refer to a change of medium (e.g. novel to film), target readership (e.g. adult to children), or any other major reworking of the text. In the present book, I follow Cattrysse (2020, 21), who argues in favour of "readopting the common parlance definition of 'translation' as the accurate rendition of the meaning of a verbal expression in another natural language, and 'adaptation' as change that leads to better fit". It must be stressed that the aim here is not to redefine translation and adaptation or discuss the implications that such redefinitions might have, but rather to simply introduce a distinction that will help highlight on the one hand the focus on denotative meaning in previous studies examining tourism translation (which would fall under the definition of 'translation' above), and on the other hand the importance of producing a tourism translation that can fulfil its promotional function (which would fall under the definition of 'adaptation' above). Adaptation, therefore, means here the process of producing a target text that takes into account the communicative preferences of the target reader when it comes to how tourism is perceived in the target language. An example of how adaptation looks like in tourism translation is provided below.

ST *Spend a day **exploring** Fort William, the Outdoor Capital of the UK and one of the most frequently used film locations in Scotland.*
TT *Passez la journée à Fort William, la capitale du plein air du Royaume-Uni et l'un des lieux de tournage les plus connus d'Écosse.*
BT *Spend the day in Fort William, the outdoor capital of the UK and one of Scotland's best-known film locations.*
Source www.visitscotland.com

In this example, the verb *explore* has been omitted, taking into consideration that French tourism discourse places less emphasis on exploration than the British English one. Such instances are considered successfully adapted, while if the verb *explorer* (*explore*) had been used, it would have been an example of lack of adaptation, and of translation based on denotative meaning alone.

3.1.1 Information Versus Promotion

A common element of many existing studies focusing on the language and translation of tourism texts is that they recognise promotion as the main function of tourism texts, and clearly state that the goal of tourism texts is to persuade (Durán Muñoz 2011; Lee 2016). Some studies adopt a more nuanced approach, however, and identify the dual function of tourism texts, arguing that while the promotional function is the goal, the informative function is the premise for this goal (Nobs Federer 1996; Sanning 2010). Although the promotional function might prevail, informative elements need to be present for it to be achieved; regarding official tourism websites in particular, Almela Soto and Navarro Coy (2018) argue that the language used aims not only to describe a destination, but also to persuade the reader-tourist to visit the place, and it is for this reason that such websites share many characteristics with advertising, where promoters try to impose their opinion on the reader-tourists. Similarly, as noted in Chapter 2, Valdeón (2009), focusing on the translation of tourism texts, describes texts produced by official tourism boards as info-promotional, whose aim is to attract visitors to an area by offering accurate information.

This dual function is achieved linguistically through certain strategies. Specifically, argumentation is used to address the logical mind, and it has mainly to do with offering relevant and accurate information, such as the distance between places, or opening hours. This is complemented by an emotional appeal, which aims to arouse interest in reader-tourists and awaken their desire to visit the place, for example by describing how beautiful or peaceful a destination is. On a linguistic micro-level, it is almost impossible to clearly differentiate between the two, and this is an additional reason why it is preferable to talk about info-promotion when it comes to most tourism texts. As already explained in Chapter 2, describing a city as *big* or a beach as *sandy* might at first sight appear as if the author is offering information about the destination, but at the same time they are highlighting characteristics of the destination that they believe are attractive to reader-tourists, while silencing those that are unlikely to attract visitors. However, as rightly observed by Nogueira da Silva (2011), these strategies do not have to be the same across languages. Different reader-tourists will have different experiences and importantly different expectations based on their previous experiences of similar texts in the target language, a point also made by Kelly (1997). In other words, the selling points that will most appeal to a specific linguistic and cultural group might be completely lost on another one.

3.1.2 Expectations of Reader-Tourists

Very little research has been conducted that focuses specifically on the expectations of reader-tourists of tourism texts regarding linguistic choices, which can be partly explained by the challenges of conducting research with focus groups and designing appropriate questionnaires that would measure expectations. Some attempts have, however, been made in recent years mainly regarding the language pair Spanish/English. For instance, Soto Almela (2013) focuses on the reception of translated tourism texts from Spanish into English by native English speakers. Specifically, he designed a questionnaire containing excerpts of the published versions of translated tourism texts produced by the official tourism board for the Murcia region, focusing on culture-specific items, and alternative versions following different translation strategies, such as domestication, foreignisation, and explicitation. The native English speakers in the study show a very strong preference for an elaborate form of explicitation, where the original item is followed by an explanation, while the published translation is deemed the most acceptable only in a small number of cases, as it does not always adopt this strategy. Importantly, the least preferred strategy is that of foreignisation, which lends support to the idea that reader-tourists, at least native speakers of English, require tourism texts to be adapted to some extent regarding culture-specific items.

Related to this is a study conducted by Nobs Federer (2005), who argues that translated tourism texts, specifically from Spanish into

English, need to conform first to a certain norm of expectation, which makes the text acceptable, and then to an adequacy norm, which makes the translation successful. However, her study remains limited to rather obvious factors that affect the acceptability and success of a text. Specifically, perhaps unsurprisingly, respondents clearly identified grammaticality and style as the most important parameters of a tourism text. This might be interpreted as evidence that adaptation is not necessary, as long as linguistic rules are followed. On the contrary, however, this is evidence of the low quality of many tourism texts, which suggests that reader-tourists often have to also lower their expectations and focus primarily on comprehensibility. Undoubtedly, before attempting to achieve anything else, a translation of any text needs to be free of linguistic errors, as its aim can be seriously compromised by lack of comprehensibility.

3.1.3 Comparing Linguistic Conventions

Focusing on tourism texts rather than their recipients, various studies have tried to compare linguistic conventions across languages. For example, Soto Almela and Navarro Coy (2018) examine how persuasion is expressed differently in Spanish and English through the use of what they term perception verbs (e.g. *explore, experience, learn, relax*). They use a comparable corpus of official tourism websites and conclude that Spanish texts rely on a wider range of perception verbs, that is, they are lexically richer, while English texts employ a small number of verbs more frequently, that is, they are more repetitive. Verbs related to cognition are preferred to those related to the senses in both languages, but English texts tend to employ more verbs such as *experience* and *explore* compared to Spanish. Spanish websites are also found to rely on verbs with more concrete meaning, which according to the authors can be seen as enriching the discourse. Specifically, *experience* is used widely in English texts to refer to perception through the senses (e.g. *experience the tastes of the UK*), while Spanish texts prefer more specific verbs related directly to the senses in similar cases (e.g. *degusta los sabores de la cocina Española—taste the flavours of Spanish cuisine*). Similarly, as far as vision is concerned, English texts show a strong preference for the verb *see*, while

Spanish texts prefer *contemplar*, which has a more specific semantic load, referring to observing carefully, especially when the environment is quiet or pleasant. In that sense, Spanish websites seem to represent tourism as an activity that implies contemplation, and tranquillity.

Another study that focuses on the comparative analysis between English and Spanish is that conducted by Martínez Motos (2005). She focuses on the investigation of a corpus of tourism brochures in English and Spanish produced by official tourism bodies in the UK and Spain, respectively. Results suggest that the Spanish brochures refer more frequently to gastronomy, architecture, art, festivals, and history compared to the English ones, which focus more on practical information than the Spanish texts. Additionally, the Spanish brochures adopt a more indirect style when addressing the reader-tourist, while the English brochures prefer a more direct one. This is clearly reflected in the preferences of English texts for imperatives (65% of clauses) and second-person pronouns (75% of personal pronouns), compared to only 5% for both in Spanish. This is in line with what both Kelly (1997) and Nobs Federer (1996) have argued in the past. At the same time, similarities were observed regarding the format of the brochures and the percentages of colloquial vocabulary used, as well as that of metaphors. Therefore, although some general similarities might be observed, there are some subtle differences between the two languages, which can be linked back to some of the observations made about the language of English tourism in the previous chapter. It is precisely because such differences are subtle, that the translation task becomes more challenging.

3.1.4 Cultural Parameters in the Study of the Language of Tourism

The studies examined so far have one feature in common, that is, they are entirely data-driven and do not rely too heavily on existing theories about cross-linguistic differences. A different approach has been adopted by Manca (2004, 2011, 2013, 2016) whose research is characterised by a clear focus on the linguistic strategies used to promote different destinations in different languages and heavily relies on the idea of cultural parameters (Brake et al. 1995; Hall 1982; Hofstede 2001; Katan 2006) as a methodological and descriptive framework. For instance, using Hall's (1982) classification of High- and Low-Context Cultures, as a measure of how explicit the messages exchanged in a culture are, Manca (2011) argues that British farmhouse holiday websites, belonging to a High-Context Culture, rely on concrete details and factual information to achieve promotion, whereas relevant Italian websites, which belong to a Low-Context Culture, rely more on descriptions using abstract nouns and aim to engage feelings.

More recently, Manca (2016) developed a more complex descriptive model for the examination of tourism discourse, which consists of three elements: (a) expressional, capturing how cultural values are expressed, (b) liaisonal, referring to how participants exchange information, and (c) structural, related to how information is structured. The first element derives from Brake et al.'s (1995) cultural orientation to Action, according to which each culture can be assigned a place on the Being-Doing continuum, as a reflection of the priorities it sets of itself relative to action. Doing cultures are said to focus on achieving goals, and improving their standards of living, and are more likely to focus on explicit messages, concrete facts, and material processes (i.e. verbs of doing such as *open*). Being cultures are said to focus on obtaining job satisfaction, improving quality of life and personal values, and rely more on implicitness, feelings, sensory language, and existential (e.g. *exist*, *remain*) and relational processes (e.g. *be*, *become*). The liaisonal element distinguishes between directness and indirectness and is associated with Hofstede's (2001) dimension of Power Distance. According to Manca, cultures favouring a more direct style of communication prefer the use of imperative forms, questions, engagement markers such as *you* and *yours*, and generally a dialogical and informal style. Conversely, cultures that prefer a more indirect style of communication are reported to use more declaratives, impersonal forms, and expressions. Finally, in Manca's model structural meaning can be divided into two components, that is, linearity and circularity, following research conducted by Katan (2006). Linearity refers to messages that are short, reader-friendly, with the main points easily identifiable, whereas circularity refers to a preference towards long and complete messages with considerable emphasis being placed on background details.

Manca then goes on to examine tourism websites in different languages, namely Italian, French, German, and English (New Zealand). Her analysis reveals that Italian and French websites demonstrate characteristics of Being cultures, with a stronger focus on indirectness and circularity, and that tourism promotion in these websites relies more on providing offers to reader-tourists, rather than making demands, as demonstrated through the low use of imperatives and directives, allowing reader-tourists to navigate through the information provided and make their own choices. On the other hand, German and English (New Zealand) websites demonstrate characteristics of Doing cultures, and rely more on linearity, with a strong focus on facts, and communicate a clear and accessible message. However, some differences are observed regarding liaison meaning, with New Zealand promotion being more direct than the German one.

As can be seen, although it is generally agreed that tourism texts require adaptation, what exactly that adaptation might involve is a question that, despite many efforts, has not so far been adequately or systematically addressed. A more in-depth understanding of the cross-linguistic differences identified will not only help to inform translation practice but also increase our awareness of how different cultures view the world differently, considering the important role that tourism texts play in identity construction (see Chapter 2). This book follows an approach similar to that of Lorés Sanz (2003) and Lee (2016), where the examination of the translation of tourism texts is guided by results obtained from a contrastive examination of the language of tourism across languages. The reason why an approach that relies on cultural parameters such as that adopted by Manca (2004, 2011, 2013, 2016) is not used here is that a theory-driven approach might result in blind spots when it comes to linguistic features that cannot be easily accommodated in the different categories relating to culture. For instance, while adjectives have been identified, at least in English, as the most typical linguistic feature of tourism texts, this linguistic feature does not feature in the cultural norms examined by Manca. It is also unclear what happens in cases where the data contradict the expectations based on the theoretical model. This is not to say that cultural norms do not affect communication, and promotion in particular, but rather that a different approach can complement the research conducted using cultural parameters. This is the approach adopted in this book, which uses the features of the tourism text as a starting point, not associating specific cultures with specific features before beginning the analysis. In other words, this means allowing texts to reveal the parameters on which these are designed, rather than imposing existing parameters which might have been designed with different products in mind. Having said that, the focus of the two case studies presented in this book is not to provide long lists of cross-linguistic differences. These differences, once identified, are interpreted through relevant theoretical concepts, namely the tourism phase and the tourist gaze, which are explained in more detail in the relevant chapters.

3.2 Previous Studies of Tourism Translation

3.2.1 Translation Errors

A large proportion of studies examining tourism translation focus on the errors identified in these texts or the (often unsuccessful) strategies employed by translators. For instance, Durán Muñoz (2011) analyses the mistakes (e.g. syntactic problems, heavy style, spelling mistakes, and mistranslations) by translators of tourism texts between English and Spanish and English and Italian. Similarly, Nogueira da Silva (2011) examines the mistakes made by translators when translating the official tourism website for Brazil from Portuguese into Spanish and identifies a number of problems including lexical choices, mistranslations, ungrammaticality, and problems related to cohesion. Finally, Fuentes Luque (2009) offers an overview of terminological problems that Spanish agrotourism texts present to translators of French and English. He concludes that translators struggle to produce successful translations due to the lack of correspondence across languages regarding the terms used for different types of accommodation. Many of the mistakes mentioned above are a result of interference and non-adherence to the communicative conventions of the target language, which, according to Durán Muñoz (2011), is explained by the fact that most of the translations are carried out by non-professional tourism translators, that is, translators who have not been adequately trained for tourism translation. Typically, the translation of tourism texts is studied and practised at universities as part of general translation, if at all. For example, according to Durán Muñoz (2012), in Spain, there are only two universities that offer specialised courses in tourism translation. However, it appears that translators are not lacking training only in tourism translation, but also in translation more generally, or more worryingly, the target language, as manifested by the numerous low-quality translations of tourism texts.

The solution which can help address these issues is often thought to be the recognition of tourism translation as a specialised type of translation, following from the idea that tourism discourse is a specialised type of discourse (see Chapter 2). Another solution would be to provide more training in tourism texts for future translators. Durán Muñoz (2011, 2012) rightly argues that translators of tourism texts need to be aware of the features that characterise the language of tourism, the anticipated challenges, and the resources available in order to meet the expectations of the target audience. She adds that translators of tourism texts need to be familiar with the specific terminology, morphosyntactic characteristics, and textual conventions of tourism texts. This goes beyond producing a simply grammatically correct text, but rather a text that can achieve its communicative purpose. However, apart from these general statements, Muñoz does not go into much more detail and does not seem to recognise that for training to be successful it needs to be informed by research into the language of tourism, which will help trainers and students identify which linguistic features need to be considered during tourism translation.

Additionally, we have to accept that a large proportion of tourism texts are likely to be translated by non-native speakers of the target languages, especially where translation is out of a less-widely spoken language. This is not to say that international promotion of certain places should cease until appropriate translators are identified, but rather that it is necessary to equip translators with the necessary knowledge, which needs to be informed by research in the linguistic features of the language of tourism across languages, to understand the expectations of reader-tourists, and to produce successful tourism texts.

Nonetheless, there are also plenty of generally error-free texts, at least as far as grammaticality is concerned, and it is this category of texts that this book focuses on. Without ignoring the detrimental effect that a low-quality translation can have on the promotional effect of a text, it is necessary to understand what happens beyond the obvious errors, some of which could have been easily corrected with editing. By focusing only on the error-infested texts, there is a risk of communicating the idea that all that is needed for destination promotion to be achieved is a grammatically correct text, when in reality much more is required. And although grammaticality is easy to fix, the more subtle nuances of intercultural communication and promotion require in-depth investigation in order to be fully grasped, and as a result improved. A thorough cross-linguistic analysis of tourism texts can provide a good starting point for such an investigation.

3.2.2 Semantic Equivalence

A large number of studies on tourism translation focus on how certain linguistic differences are dealt with in translation. Such linguistic differences typically include terminology and culture-specific items, which are often easy to identify and analyse. The focus is on whether and how semantic equivalence can be achieved and the strategies that translators employ in that regard. For instance, regarding terminology, Fijo León et al. (2013) analyse terms related to rural tourism accommodation establishments in three languages (Spanish, English, and French) to identify similarities and differences across languages, with the aim of suggesting appropriate translation equivalents. They conclude that although all three languages share the generic concept of rural tourism accommodation establishments, the terms used to represent this concept vary significantly, with the result that one-to-one correspondence in translation can rarely be found, which inevitably results in translation challenges. The existence of such terminological challenges is perhaps further indication that tourism forms a specialised type of discourse.

Culture-specific items in tourism texts, whether these refer to material culture, social culture, ecology, or the arts, are another popular linguistic feature when examining tourism translation. A study conducted by Almela Soto (2013) focuses on a range of Spanish culture-specific items, including history, traditions, and gastronomy and asks native speakers of English their preferences regarding different strategies when translating these items. Results suggest that explicitation is the preferred translation strategy, and that respondents considered the published translation as 'better' than that provided by the researcher only in a small number of instances (approximately 30%). As a result, Almela Soto concludes that published tourism texts do not manage to fulfil their function successfully. Another study focusing on culture-specific items is that conducted by Terestyényi (2011), which is worth mentioning here as it focuses on Hungarian into English translations of tourism texts, which are likely to have been performed by non-native speakers of English. Like Almela Soto, Terestyényi focuses on a number of culture-specific items, including in the areas of geography, gastronomy, traditions, art, and culture; she examines the techniques used to translate these items in official travel brochures. She concludes that a number of techniques are employed, with transference being the most popular one, and argues that the strangeness that this technique introduces to the text contributes to arousing interest in potential visitors. While this might be true, it does not take into account the expectations of reader-tourists, who seem to prefer such terms to be explained according to Almela Soto (2013). This further stresses the need for a better understanding of key linguistic features of tourism texts across languages, also taking into account the expectations of reader-tourists.

Some studies aim to go a step beyond the simple examination of some of the most obvious anticipated lexical challenges in the language of tourism and how these have been dealt with in translation. These include studies exploring collocations, such as those conducted by Manca (2004, 2008), who examines the phraseology of British and Italian agrotourism websites focusing specifically on linguistic items such as *position*, *landscape*, *cottage*, and *nature*. She follows an approach based on which she first examines the lexico-grammatical profile of a word to identify its most frequent collocates in the corpus of agritourism websites, and for each word, a prima-facie translation is identified which is then examined for collocates. Then, for each prima-facie translation, the most successful translation equivalent is identified for the initial search word. Although her studies aim to capture mistranslations, therefore also including errors, she links collocational patterns to promotional strategies, arguing that a systematic study of these can help translators produce texts that can better achieve their function. Importantly, Manca rightly observes that "function is very difficult to pin down and even more difficult to transfer across languages" (2004, 38). This focus on function is of significance, as it marks a rather belated recognition that tourism texts can have very different source and target audiences, with their own cultural expectations, and that translation strategies examined without reference to respective non-translated texts in the target language, which can offer valuable cross-linguistic insights, are of limited value to both translation theory and practice.

3.2.3 Translating Promotion

Some studies try to move away from specific linguistic features, and instead focus on the function of tourism texts, that is, mainly promotion, to understand whether and to what extent tourism texts successfully achieve this promotional function. Although techniques of psychological evaluation might be more suitable for testing the addressees' perception of and reaction to translations to examine whether and to what extent promotion is achieved, there is a lot to be gained from linguistic analyses, which will hopefully in the future inform more research into reader-tourists' reaction to texts.

A strong emphasis on the promotional function of tourism texts can be found in studies focusing on the pragmatic meaning of tourism texts. One of the most insightful examinations is that offered by Hickey (2004) who clearly situates his study within the field of pragmatics and aims to identify how perlocutionary effect, that is persuading reader-tourists to visit a destination, is achieved through translation. He rightly observes that if translators aim at semantic equivalence only, this may not bring about the same reaction in reader-tourists from different linguistic and cultural backgrounds. He examines a Spanish brochure and its translation into English and French in terms of a number of features, such as superlatives, hedges, assertions, and statement of facts. He concludes that there is little evidence to suggest that the translators have aimed for perlocutionary equivalence, instead semantic equivalence tends to be prioritised. He seems to put the blame on translators, who, he argues, do not recognise that the aim of these texts is to achieve promotion, but it remains questionable whether or not translators actually know how promotion is achieved in these texts in terms of both the source and target language. Hickey's (2004) small-scale study is probably the only one to focus so closely on pragmatic meaning, which makes a notable addition to the body of studies on tourism translation. However, his observations about whether or not perlocutionary equivalence is achieved are based on subjective observations and not on the comparable analysis of texts. A similar attempt is made by Hatim (2004) who, however, finds more merit in the attempt of the translator to transfer Arabic tourism texts into English.

A similar approach with a focus on the function of tourism texts, but with less of an emphasis on pragmatics, has been adopted by numerous other scholars. For instance, Valdeón (2009) investigates how English translations of Spanish texts deal with a number of linguistic features, but limits himself to the obvious candidates of grammar and culture-specific items. However, he identifies problems with the translations that are related to pragmatics, such as grammatically correct forms which nevertheless do not take into account the shift of perspective regarding the audience. Importantly, he stresses the need for translators to consider the context and the characteristics of their new audience in order to produce successful translations and calls for an interdisciplinary collaboration between cultural studies, discourse analysis, and translation. Likewise, Agorni (2012) examines how a wide range of linguistic features, including similes, metaphors, colloquial language, rhetorical questions, and evaluative language, which she associates with promotion, are treated in translation from Italian into English. Although not an in-depth study, she concludes that translators linguistically manipulate the target text showing evidence that they are sensitive to these features and the overall effect that their texts might have on reader-tourists. Although this recognition is important, she seems to imply that it is up to the translators to decide how to manipulate the text successfully, ignoring the fact that they need to take the reader-tourists' expectations into account and decide how these can be met in the TT. Finally, Sulaiman (2014) examines the English and Malay versions of the official tourism website for Australia. He focuses on the stylistic features of these texts and identifies imperatives, among other linguistic features, as one of the features characterising English texts. However, when translated literally into Malay these imperatives have a negative impact on the promotional function of the text, similar to Valdeón's (2009) observations about grammatically correct, but pragmatically unsuccessful translation choices. He argues that an indirect storytelling style would have been more appropriate for the target audience.

All the studies discussed above, like those of Hickey (2004) and Hatim (2004), rely on subjective statements about the success of the translations in their study and do not examine non-translated texts in the target language, with the exception of Sulaiman (2014) who relies on a brief examination of non-translations in Malay. Sulaiman also makes an important contribution to the field of tourism translation by recognising that tourism texts need to have an 'overall stylistic effect', which is the result of a combination of devices and features, and arguing that translators should not follow a strict approach where, for instance, all direct forms of expression are converted to indirect ones, as long as the overall style and feel of the target text point towards indirectness. Therefore, close readings of source texts and their translations, which is standard practice in research into tourism translation, are unlikely to paint a complete picture of the phenomenon under investigation. What is needed instead is a holistic approach to those texts, focusing on their overall effect, which will also be adopted in the case studies that form part of this book. More specifically, instead of examining how, for example, each direct form of address has been translated, the total number of direct and indirect forms can be counted to establish the overall effect of the text.

There are also examples of studies into the function of tourism texts and their translations which rely on detailed examinations of non-translated texts in the target language to support the comparison between source and target texts. For instance, Lorés Sanz (2003) focuses on the analysis of conjunctions and discourse markers in what she calls the 'thematic position' in Spanish and English tourism texts (e.g. travel guides, brochures, etc.). By examining a comparable corpus of non-translated texts in both languages, she found that Spanish texts, irrespective of the category they belong to, make a much higher use of connectors (such as *además* and *también*, both meaning *also*) compared to English texts, showing a tendency towards making logical relations more explicit. Based on these findings, it might be expected that English translations of Spanish texts need to adapt, that is lower the frequency of connectors in thematic position if these texts are to meet the expectations of the target audience. However, the examination of a parallel corpus reveals that translations exhibit very similar patterns to those found in the Spanish source texts, decreasing the use of connectors only very slightly. This means, according to Lorés Sanz (2003), that English reader-tourists of the translation might feel that they need to make less effort to comprehend the text, or, alternatively, they might dislike this interference. Although she concludes that the English translations do not follow the communicative conventions of the target language, she does notice a degree of negotiation between the source and target language, as demonstrated by the slightly lower number of connectors in thematic position found in the translations compared to the source texts. She interprets this as an indication that the target texts in her study sit between the source and the target language; they are an example of the third code (Frawley 1984). As a result, she argues, rather oddly, that we need to accept translated tourism as a genre with its own conventions.

However, the evidence of negotiation that Lorés Sanz (2003) provides is rather scant. Specifically, examining the total number of connectors across tourism texts, these appear with a frequency of 13.7% out of the total number of words in Spanish non-translated texts, 2.1% in English non-translated texts, and 11.6% in English translated texts. The difference between source and target texts is small, namely 16.6%, and it is not statistically significant, something that Lorés Sanz however does not comment on or take into consideration when drawing conclusions. At the same time, the distance from the communicative conventions of non-translated English texts is considerable. For these reasons, the conclusions reached risk creating an impression that translations of tourism texts are not expected to follow specific linguistic conventions and that there is no benchmark against which they can be compared. Although Lorés Sanz contemplates how reader-tourists might react to these texts, she does not come to any conclusion based on her observations. However, for a genre that is so tightly linked to how we perceive the world around us and which also has the potential to drive an entire industry, we cannot grow complacent and accept translations that do not meet the target audience's expectations as simply a genre in their own right. This is because reader-tourists are unlikely to distinguish between translated and non-translated texts as different genres, and therefore will have the same expectations from both.

A similar study is that conducted by Lee (2016), who examines the distribution of what he calls "theme types and interactional themes" in Korean and English official tourism websites. Results from his comparable corpus suggest that Korean and English tourism texts differ significantly regarding thematic patterns. Specifically, although English texts occasionally rely on process themes (e.g. *enjoy*, *explore*, etc.), these are next to impossible in Korean, according to Lee. Conversely, Korean texts make a much higher use of circumstance themes (e.g. *in the summer*, *on the island*, etc.) than English. Some subtle differences are also observed regarding existential themes (e.g. *there is*) which, although rare in the English texts, do not exist in Korean. Finally, it was found that English tourism texts rely much more on interactional themes (e.g. *you*, *we*, imperatives, and interrogatives) than Korean texts. Lee (2016) hypothesises that English translations will employ process themes less frequently and circumstance themes more frequently than English non-translated texts, as a result of staying close to patterns found in the Korean source texts, although it is not clear why interference from Korean source texts is expected. Similarly, he hypothesises that translations will employ fewer interactional themes than respective non-translated texts. Results from the parallel corpus suggest that both hypotheses are supported, but the phenomenon of the third code is also observed here. Specifically, English translations stand somewhere between Korean source texts and English non-translated texts, while they are sufficiently far away from Korean texts to be distinguished from them. Ultimately, they fall critically short of resembling comparable English texts. Similar to the conclusion reached by Lorés Sanz (2003), the target text shows evidence of compromise or negotiation between the source and target language.

However, unlike Lorés Sanz (2003), Lee (2016) does not accept this as simply a feature of translated texts. For instance, regarding imperatives as interactional themes, 261 were found in English non-translated texts, 16 in translated ones, and only 4 in Korean texts (each corpus consisted of approximately 30,000 words). This indicates that translators did not make an attempt to introduce additional imperative clauses into the target text. This might be because such shifts are considered substantial and require significant cognitive effort and creativity. However, as rightly observed by Lee (2016), such differences have a serious impact on the function of the target text. Having a target text that is almost entirely composed of declarative sentences makes it appear boring and uninviting to the English-language reader-tourists who are used to a different communicative style, therefore compromising its promotional function.

Achieving semantic equivalence might be a fairly simple thing to achieve (even though the numerous examples of low-quality tourism translations might suggest otherwise), but this is often at the expense of perlocutionary equivalence. The translator should be in a position to understand how perlocution is achieved linguistically, which can be achieved firstly through understanding what this effect means in tourism discourse and secondly through knowledge of the differences in communicative conventions between source and target language, to prevent the source text from almost imposing itself on the target one. This might require radical changes to the target text to create a text that has sufficient appeal to its target audience. We are only now beginning to develop a certain degree of sensitivity towards the importance pragmatic equivalence merits in translation of tourism texts, and therefore more in-depth studies are required which can inform both translation theory and the training of future translators of tourism texts.

3.3 Methodological Considerations

3.3.1 Qualitative Studies

Methodologically, previous studies belong to two broad categories: qualitative studies and quantitative studies. The first category of studies is interested in the examination of non-numerical data; it does not focus on counting how frequently certain linguistic phenomena make their appearance in translated and non-translated texts, but rather on how exactly a source fragment is rendered in the target language. For that reason, they can focus on a wide range of features, typically within the same study, from the elusive concept of style to terminology and proper nouns. The earliest examples of such qualitative studies include the comparison of source and target texts and the discussion of selected examples, such as that offered by Kelly (1997), who presents data from Spanish tourism texts translated into English by discussing randomly selected examples, some of which are paragraph-long, while others consist only of a few words. Similarly, Pierini (2007) presents selected examples of English translations of Italian tourism texts, some of which are fairly long. A similar approach is adopted by many other scholars (e.g. Durán Muñoz 2011). These studies typically focus on translation errors and strategies, and they are rarely concerned with establishing the differences in communicative conventions between the source and the target language prior to the analysis of parallel texts. In other words, a comparable analysis of non-translated tourism texts in the two languages, which would help establish whether and what type of differences can be observed, has not been conducted before examining how specific linguistic items have been translated.

One of the first attempts to go a step further is Valdeón's (2009) study, which presents first a discussion of the features of the Spanish source texts before examining how they have been transferred into English. While this additional stage of analysis is certainly useful, it does not add very much to the discussion because a similar stage examining the features on non-translated English texts, which would have helped establish the differences between Spanish and English, is missing. More significant from a methodological point of view is the study conducted by Sulaiman

(2014), which was discussed earlier. The additional stage of analysis based on the features of non-translated texts in both source and target languages allows the researcher to put forward a more convincing argument about where exactly the target texts stand compared to the source text and respective non-translated texts in the target language.

Overall, it is perhaps important to understand that qualitative studies were born out of a desire to comment on and analyse the low quality of tourism translation. For this, what is often required is no more than a handful of examples of bad translations. However, as the interest in tourism translation grew, more systematic studies started to emerge, which also tried to analyse non-translated texts. Although the examples provided are often limited to allow for a comprehensive picture to be built, including a contrastive analysis of non-translated tourism texts makes argumentation more convincing and moves tourism translation research into new and more interesting areas of investigation.

3.3.2 Quantitative Studies

The second category of studies, which also tend to be more recent, relies on quantitative data, mainly through the use of corpora, for the investigation of the language of translation. Corpora are understood here as large electronic collections of texts which have been compiled according to specific design criteria and which can be analysed (semi-) automatically using a range of corpus tools created for linguistic research. Corpus-based studies focusing on the translation of tourism texts typically incorporate both a comparable and a parallel component, with contrastive studies of non-translated texts in two or more languages preceding the examination of translated texts, although there are some exceptions to this pattern. For example, Cappelli (2008) compiled two subcorpora of websites promoting Siena: one Italian-English comparable corpus of non-translated texts and one Italian-English parallel corpus of source and target texts. The corpus is fairly small, around 300,000 words, and is used to conduct a number of keyword analyses, that is, words that are unusually frequent compared to the British National Corpus. Similarly, Manca (2008) uses a comparable subcorpus of Italian and British agrotourism websites, and a parallel subcorpus of Italian source texts and their English translations. Overall, the corpus is one of the biggest ever compiled focusing on tourism texts at approximately 1.5 million words. However, there is a certain imbalance in the design, with the parallel component only reaching 100,000 words. Finally, as already discussed, a combination of comparable and parallel corpora is also used by Lorés Sanz (2003) and Lee (2016). In the case of Lorés Sanz (2003), data consists of a corpus of tourism texts which is divided into three components: a component of Spanish texts, a component of English texts, and a component of Spanish texts translated into English. The first two components are combined to create a comparable corpus, which allows the researcher to establish whether there are any differences in the use of connectors between the two languages. Then, the first and third components are combined to create a parallel corpus to investigate how these differences are treated in translation. No information is provided on the size of the corpus, but if we consider that tourism texts are rarely very lengthy, and that altogether 21 texts were used, it is safe to assume that the corpus is probably fairly small. Similarly, Lee (2016) examines data in three types of texts taken from official tourism websites of different cities and provinces in Korea and the US: Korean non-translated, English non-translated, and English translations of the Korean texts. However, once again, the total corpus is rather small at approximately 100,000 words.

The reason for the rather small corpus size can be explained to some extent by the specialised nature of tourism texts; it is widely accepted that corpora capturing specialised language tend to be smaller than general language corpora. However, considering the proliferation of tourism texts, and in particular tourism websites, in recent years, this very small corpus size, especially as far as the parallel subcorpus is concerned, is not easy to justify. Nevertheless, what is particularly interesting about these corpus-based studies is that they combine both comparable and parallel investigations—a clear step towards triangulation, which can be understood here as the combination of multiple corpus values (i.e. from the category of corpus type) (Malamatidou 2018). This is to be expected since the comparable subcorpus is necessary in order to establish communicative conventions and therefore reader-tourists' expectations, which can then inform translation training and practice. However, triangulation is currently limited to this combination, while other possibilities remain unexploited, for example the combination of multiple languages attributes, that is, data from more than two languages. The only possible exception to this is the corpus-based study conducted by Fijo León et al. (2013) which compared non-translated texts in English, Spanish, and French. However, since a parallel subcorpus was not included in this study, it still misses the opportunity to become a more comprehensive examination of the language of tourism translation.

Methodologically, it is also worth examining the specific quantitative methods employed in these studies. In the majority of cases, simple frequency counts are used together with percentages or other normalised frequencies. However, relying on frequency counts only hides the risk of misinterpreting even small percentage differences for meaningful differences between languages and texts. The only study where a combination of descriptive and inferential statistics is successfully achieved is that conducted by Lee (2016), who apart from raw and normalised frequencies also calculates the probability value p to establish whether differences observed are due to chance, and odd ratios to measure the strength of association. In that respect, although relying on a small corpus size, Lee's (2016) is the most comprehensive quantitative study from a methodological point of view, and manages to achieve corpus method triangulation, which refers to the combination of two or more corpus analysis techniques (Malamatidou 2018). This engagement with multiple methods for the analysis of data is just beginning to attain the importance it merits in both tourism translation studies and translation studies more generally.

3.4 Taking Stock

The examination of previous studies focusing on the translation of tourism texts has helped reveal some of the main characteristics of this area of research, as well as its main limitations. Although a number of studies are available, this number is not representative of the importance of tourism in modern economies and the consequent proliferation of translated tourism texts. Additionally, perhaps not surprisingly, the majority of studies focus on either Italian or Spanish and English, that is languages of countries in Europe with a strong tourism industry. However, there are many more countries that have either a strong or an emerging tourism industry which have been neglected so far. Importantly, very few of these studies are conducted from the perspective of English, that is, how English tourism texts are translated into other languages. Yet, many English-speaking destinations, such as the USA, UK, and Australia, are very popular among international tourists, and therefore it is important to turn our attention to what happens when the source text is English. In addition to this, a recent report by Visit Britain (2019), the country's national tourism agency, responsible for marketing Britain worldwide and developing Britain's visitor economy, reported that only approximately 40% of visitors felt their expectations were met regarding the signs (e.g. road signs) and information available in their language. Although this covers much more than official tourism websites, it is obvious that linguistic considerations are important and can affect visitor satisfaction. Therefore, more studies are needed which focus on a wider range of languages, and if possible, allow for cross-comparisons among these.

Another key characteristic of previous studies, whether qualitative or quantitative, is that they rely on small amounts of data. In the case of qualitative studies, these are randomly selected examples, typically chosen because they help support the author's argument. Therefore, although they can offer valuable insights, it is often difficult to judge whether these studies are truly data-driven. Conversely, quantitative studies use only small pockets of data, which make generalisations problematic. Having said that, a number of quantitative studies, although focusing on different languages and linguistic features, seem to agree that target texts in their studies stand somewhere between source texts and respective non-translated texts in the target language, although the balance leans more towards the source texts. An empirical large-scale study is yet to be conducted; this would provide a comprehensive answer to how linguistic differences and reader-tourists' expectations are negotiated during translation.

Furthermore, the focus of previous studies is on a range of linguistic features, whether that is to highlight translation errors, examine translation strategies, or more generally explore the nature of tourism translation. Studies that focus on pragmatic properties tend to offer more significant insight into the phenomenon than those focusing on how culture-specific items are being translated. This is not to say that the latter is not of interest, but rather, if we are to foreground translation as a space where cross-linguistic differences contributing towards the promotional function of texts can be negotiated, we need a much deeper understanding of how different, and less conspicuous, linguistic features are treated in translation. Related to this point, even studies focusing on pragmatic features ultimately reach a conclusion that has a vocational element to it. In other words, the aim is to understand what happens in translation in order to make qualitative judgements and inform translation training and practice. This is a worthy aim, of course, as there is no denying that tourism texts would benefit from a more informed approach to translation. However, this vocational focus seems to be the only aim of previous studies, which tend to neglect the potential that tourism texts offer for understanding how specific linguistic choices are shaping behaviours and expectations of reader-tourists, ultimately revealing how different cultures do indeed view the world differently. Tourism texts, in addition to their important instrumental function, which is to turn readers into visitors, also offer a window into how we perceive the world, which elements of it we prioritise, and why and how we relate to it through language. Social reality shapes and is being shaped by tourism texts, but as these texts travel from one language to another, this reality shifts, which is also partly why reader-tourists' expectations change. However, this connection between tourism texts and our perception of the world has not been touched upon in previous studies focusing on the translation of tourism texts.

Related to this last point, what is particularly surprising about previous studies is that they almost completely ignore research into tourism itself. With the exception of occasional references to Dann (1996) and his research on the linguistic features of tourism texts, there is very little interest regarding how a destination image is built via tourism texts from a marketing point of view, or any other relevant considerations, such as how personal factors might affect the language employed in tourism. However, without identifying those aspects that play a key role in tourism, previous studies are risking being reduced to linguistic exercises, which, although useful, remain largely relevant only to translation studies and miss the opportunity to engage in a meaningful dialogue with the tourism industry, when, after all, the translation brief comes from destination promoters. Therefore, they miss a valuable opportunity to truly transform tourism translation.

3.5 A Novel Approach to Tourism Translation Research

Previous studies on the translation of tourism texts can be viewed here as the important first step towards a better understanding of how tourism is adapted cross-linguistically. By examining their main characteristics and limitations, it is apparent that a large-scale empirical, interdisciplinary study that analyses a number of languages and language pairs (e.g. both from and into English and from and into languages other than English), that focuses on linguistic features pertinent to elements of the destination image and that are informed by research conducted in tourism studies, will go beyond a vocational approach to translation. Ultimately, such a study has the potential to create positive implications for both the translation and tourism industries, as well as for our understanding of how we perceive the world around us. This final section of the chapter will explain how exactly this approach has been developed in the current study.

A distinct aspect of the approach in the present book is that it is interdisciplinary, allowing translation studies to stretch its boundaries and find a long overdue relevance in new disciplines and practices. It relies on ideas associated with tourism destination promotion as a starting point, and, as a result, is solidly grounded in research conducted in the tourism industry. Therefore, the linguistic features studied are not randomly selected or based on comparative research in the languages examined; rather, they reflect elements that are indispensable for tourism to take place. These are the tourist and the destination. This book will therefore focus on these two elements, which will be expanded into two case studies: the tourist behaviour (realised in 'things to do') and the physical space (realised in 'places to go'). The rationale for these case studies, as well as how they complement and occasionally overlap with each other, will be discussed in the corresponding chapters. These explorations will help reveal the complexity of destination promotion through written texts, the cross-linguistic differences, and the role that translation plays in negotiating these. Having a solid tourism research background, this study is then able to move beyond the vocational focus that characterises past research, and in particular explore the way in which tourism texts encode how we experience and view the world. The way we are used to perceiving the world, and the way this is presented to us, is tightly linked to where we come from. If we examine the narratives that different languages use to describe tourism destinations, we can gain some insight into how readers experience tourism and how they gaze at the world around them, for instance which elements they pay particular attention to. Ultimately, the aim is to examine what role translation plays in negotiating any differences observed.

Similarly, tourism is also considered to promote identity creation for the destination and its host, as well as for the visitor. This is to be understood at the level of physical space, as geographical locations reconfigure themselves to become tourism destinations (Hughes 1992, 1998; Short et al. 1993). While many processes are set in motion to achieve this, a key role is played by the tourism narrative, which is what transforms, say, an island into the ideal destination for summer holidays. It is therefore not surprising to hear stories of people rushing to a destination, only after it has been branded as a place that must be visited. However, this process of identity creation is highly collaborative and is negotiated in interaction with the potential visitor. Destination promoters, in order to construct the identity of a place as a tourism destination, are likely to consider their market of potential visitors and adjust the Tourism Destination Image (TDI) accordingly in terms of what these are looking for, a process which is also likely to be affected by current trends and habits as promoted by the media (e.g. 'instagramability' as one of the most recent trends). At the same time, this interactive process also contributes to the construction of the visitor's identity (Hallett and Kaplan-Weinger 2010). Specifically, by interpreting and consuming tourism texts, and later by visiting destinations, reader-tourists engage in a process of constructing themselves as tourists to particular destinations. What is more, tourism texts, through the linguistic and other means they employ, form expectations and make promises. As a result, when ultimately visiting a destination, it can be hypothesised that visitors are likely to condition their behaviour to match these expectations, creating an identity of self as tourist which is the result of the previous interaction with the discourse of tourism. It is clear that this process of identity construction for both the destination and the tourist has a certain element of circularity, with the tourist both feeding into it and being fed by it. This central role that the tourist plays needs to be taken into consideration when considering tourism texts which incorporate a bilingual or multilingual element to examine whether these texts construct the same or different identities across languages and if so how, something which has not been attempted so far. Having said that, it is also possible of course that the importance of destination branding and/or identity construction is not recognised during the production of the tourism text, which can result in it not achieving its desired objective.

In summary, the main aim of this novel approach to tourism translation is to determine the crucial link between the micro-level textual practices and macro-level socio-cultural and promotional processes. In this regard, language is not seen as a static, materialistic component, but as a dynamic, meaningful element shaping and being shaped by considerations of image formation. The role of translation is examined in detail, as it plays a key role in potentially strengthening or breaking this link by creating new images and identities or failing to do so. The rest of the book focuses on the insights that can be gained when applying this approach to empirical data.

References

Agorni, Mirella. 2012. "Tourism Communication: The Translator's Responsibility in the Translation of Cultural Difference." *PASOS: Revista de Turismo y Patrominio Cultural* 10 (4): 5–11.

Almela Soto, J. 2013. "La Traducción de Términos Culturales En El Contexto Turístico Español- Inglés: Recepción Real En Usuarios Anglófonos [The Translation of Cultural Terms in Spanish-English Tourism Context: Actual Reception of Anglophone Users]." *Quaderns* 20: 235–50.

Almela Soto, J., and M. Navarro Coy. 2018. "Los Verbos de Percepción En El Discurso Turístico Promocional: Estudio Contrastivo Inglés/EEspañol [Perception Verbs in Tourist Promotional Texts: A Contrastive Study of English/Spanish]." *Babel* 64 (5–6): 649–70.

Brake, T., D. Medina-Walker, and T. Walker. 1995. *Doing Business Internationally: The Guide to Cross-Cultural Success*. Burr Ridge, Illinois: Irwin.

Cappelli, Gloria. 2008. "The Translation of Tourism-Related Websites and Localization: Problems and Perspectives." In *Voices on Translation*, edited by A. Baicchi, 97–115. Rome: Bulzoni Editore.

Cattrysse, Patrick. 2020. "Translation and Adaptation Studies: More interdisciplinary reflections on theories of definition and categorization." *TTR* 33 (1): 21–53.

Dann, Graham M. S. 1996. *The Language of Tourism*. Oxon: CAB International.

Durán Muñoz, I. 2011. "Tourist Translations as a Mediation Tool: Misunderstandings and Difficulties." *Cadernos de Tradução* 1 (27): 29–49.

———. 2012. "Necesidades de Mejora y Adecuación En La Traducción de Textos Turísticos Promocionales [Improvement and Adequacy Needs in Translation of Promotional Tourist Texts]." *Hermeneus* 14: 263–78.

Fijo León, Maria Isabel, and Adrian Fuentes Luque. 2013. "A Corpus-Based Approach to the Compilation, Analysis, and Translation of Rural Tourism Terms." *Meta: Journal Des Traducteurs* 581: 212–26.

Frawley, William. 1984. "Prolegomenon to a Theory of Translation." In *Translation: Literary, Linguistic, and Philological Perspectives*, edited by William Frawley, 159–75. London and Toronto: Associated University Press.

Fuentes Luque, A. 2009. "El Turismo Rural En España: Terminología y Problemas de Traducción [Rural Tourism in Spain: Terminology and Translation Problems]." *Entreculturas* 1: 469–86.

Hall, E. 1982. *The Hidden Dimension*. New York: Doubleday.

Hallett, R., and J. Kaplan-Weinger. 2010. *Official Tourism Websites: A Discourse Analysis Perspective*. Bristol: Channel View Publications.

Hatim, B. 2004. "Culture as Textual Practices: The Translation of the Tourist Brochure as Genre." In *Pragmatics at Work: The Translation of Tourist Literature*, edited by M. Navarro Errasti, R. Lorés Sanz, and S. Murillo Ornat, 15–27. Bern and Oxford: Peter Lang.

Hickey, L. 2004. "Perlocutionary Pursuits: Persuading of/That/To." In *Pragmatics at Work: The Translation of Tourist Literature*, edited by E. Navarro, R. Lorés Sanz, and S. Murillo Ornat, 57–79. Bern and Oxford: Peter Lang.

Hofstede, G. 2001. *Culture's Consequences: Comparing Values, Behaviors, Institutions, and Organizations*. Thousand Oaks, CA: SAGE Publications.

Hughes, George. 1992. "Tourism and the Geographical Imagination." *Leisure Studies* 11: 31–42.

———. 1998. "Tourism and the Semiological Realization of Space." In *Destinations: Cultural Landscapes of Tourism*, edited by Greg Ringer, 17–32. London and New York: Routledge.

Katan, D. 2006. "It's a Question of Life or Death: Cultural Differences in Advertising Private Pensions." In *Forms of Promotion: Texts, Contexts and Cultures*, edited by N. Vasta, 55–80. Bologna: Pàtron Editore Bologna.

Kelly, Dorothy. 1997. "The Translation of Texts from the Tourist Sector: Textual Conventions, Cultural Distance and Other Constrains." *TRANS* 2: 33–42.

Lee, C. 2016. "A Corpus-Based Investigation of Theme Choice in English Translations of Korean Online Tourist Texts—With Focus on Interactional Themes." *Perspectives: Studies in Translatology* 24 (2): 294–318.

Lorés Sanz, Rosa. 2003. "The Translation of Tourist Literature: The Case of Connectors." *Multilingua* 22: 291–308.

Malamatidou, Sofia. 2018. *Corpus Triangulation: Combining Data and Methods in Corpus-Based Translation Studies*. London and New York: Routledge.

Manca, E. 2004. *Persuasion in Tourism Discourse: Methodologies and Models*. Cambridge: Cambridge Scholars Publishing.

———. 2008. "From Phraseology to Culture: Qualifying Adjectives in the Language of Tourism." *International Journal of Corpus Linguistics* 13 (3): 368–85.

———. 2011. "The Analysis of Meaning Between Language and Culture in the Tourism Domain." *Languages in Contrast* 11 (2): 172–92.

———. 2013. "Describing Through the Five Senses: A Contrastive Socio-Cultural and Linguistic Analysis of Italian and British Tourist Websites." In *Tourism and Tourist Promotion Around the World*, edited by E. Manca and F. Bianchi, 109–24. Lecce, Italy: Università del Salento.

———. 2016. "Official Tourist Websites and the Cultural Communication Grammar Model: Analysing Language, Visuals, and Cultural Features." *Cultus: The Journal of Intercultural Mediation and Communication* 9 (1): 2–22.

Martínez Motos, R. 2005. "Análisis Comparativo de Las Convenciones Textuales de Textos Turísticos En Inglés y En Español: Los Folletos Editados Por Organismos Oficiales [Comparative Analysis of the Textual Conventions of Tourist Texts in English and Spanish: Brochures Edited by Of]." In *La Traducción En El Sector Turístico* [Translation in the Tourist Industry], edited by A. Fuentes Luque, 121–38. Granada: Editorial Atrio.
Nobs Federer, M. 1996. "Reflexiones Sobre La Problemática Específica de La Traducción de Textos Procedentes Del Campo Del Turismo [Reflections on the Specific Problem of Translation of Texts from the Field of Tourism]." In *Perspectivas Hispanas y Rusas Sobre La Traducción: Actas Del II Seminario Hispano- Ruso de Estudios de Traducción, Universidad de Granada, 3-7 de Abril de 1995* [Perspectives on Spanish and Russian Translation: Proceedings from the II Seminar on Spanish-Ru], edited by E. Iniesta, 239–53. Granada: University of Granada.
———. 2005. *La Traducción de Folletos Turísticos: ¿Qué Qualidad Demandan Los Turistas?* [The Translation of Tourist Leaflets: What Quality Do Tourists Require?]. Granada: Comares.
Nogueira da Silva, A. 2011. "Análisis de La Traducción al Español de La Página Web Del Portal Del Ministerio Brasileño de Turismo [Analysis of the Spanish Translation of the Website Portal of the Brazilian Ministry of Tourism]." *Cadernos de Tradução* 1 (27): 51–70.
Pierini, Patrzia. 2007. "Quality in Web Translation: An Investigation into UK and Italian Tourism Web Sites." *The Journal of Specialised Translation* (8): 85–103.
Sanning, H. 2010. "Lost and Found in Translating Tourist Texts: Domesticating, Foreignising or Neutralising Approach." *JoSTans* 13: 124–37.
Short, J.R., L.M. Benton, W.B. Luce, and J. Walton. 1993. "Reconstructing the Image of an Industrial City." *Annals of the Association of American Geographers* 83 (2): 207–24.
Sulaiman, Mohamed Zain. 2014. "Translating the Style of Tourism Promotional Discourse: A Cross Cultural into Stylescapes." *Procedia—Social and Behavioral Sciences* 118: 503–10.
Terestyényi, E. 2011. "Translating Culture-Specific Items in Tourism Brochures." *Skase Journal of Translation and Interpretation* 5 (2): 13–22.
Valdeón, Roberto A. 2009. "Info-Promotional Material Discourse and Its Translation: The Case of the Asturias Tourist Board Texts." *Across Languages and Cultures* 10 (1): 21–47.
Visit Britain. 2019. "The Visitor Experience in Britain: Welcome, Expectations, Satisfaction and Recommendation."

4

TrAIL Corpus and General Methodology

This chapter introduces the data that will be analysed as part of the two case studies in this book, as well as some general considerations around the methodology employed for their analysis. The specific methodology used in each case study, that is how verbs and adjectives are identified and the semantic categories used for their classification, will be presented in detail in Chapters 5 and 8, respectively. As already explained, the most comprehensive studies of tourism translation employ corpus-based methods and seem to rely, at least to some extent, on the principle of triangulation (see Chapter 3). This is also the approach adopted here: a large electronic collection of texts, in this case official tourism websites, compiled according to specific design criteria will be analysed semi-automatically using a suite of corpus tools. More specifically, corpus data and method triangulation (Malamatidou 2018) is employed, which refers to the combination, in an integrated manner, of multiple values and attributes from one or more corpus variables, and the use of two or more corpus analysis techniques in the study of a single phenomenon.

Like some previous studies, this study also relies on two main stages of analysis: a contrastive/comparable one, where non-translated texts in different languages are compared, and a translation/parallel one,

S. Malamatidou, *Translating Tourism*, Palgrave Studies in Translating and Interpreting, https://doi.org/10.1007/978-3-031-49349-2_4

where the source texts and their translations are compared. This design also allows for a third stage: a comparison between translated and non-translated texts. Therefore, although the focus is ultimately on translation, in order to understand how exactly differences between languages are negotiated in translated texts, it is important to first establish what these differences entail. This is seen as a preparatory stage, informing the exploration of translation. At the same time, following Lee (2016), both descriptive and inferential statistics, which will be explained in detail below, will be used to analyse corpus data. Working with a large corpus and conducting multiple comparisons across language pairs will give rise to a large number of numerical data sets, not all of which are likely to be significant. Corpus method triangulation will help identify those patterns that are significant and meaningful.

4.1 TrAIL Corpus

4.1.1 Official Tourism Websites

The two case studies presented in this book rely on the examination of a purposefully built corpus, which consists of official tourism websites promoting the USA, the UK, France, Greece, and Russia. As already discussed in Chapter 2, official tourism websites offer rich multilingual data, being translated often into many languages (although see also discussion below), while they also place much greater emphasis on the Tourism Destination Image than other types of tourism texts. They represent the official narrative of the destination about how it is to be perceived by both the local population and the rest of the world. As such, these web-based texts are not only promotional, but to a large extent ideological. Due to their significant potential for shaping perceptions about the destination, they have been chosen as the focus of the present book. The corpus is named TrAIL, which stands for *Translation Across and In-between Languages* and reflects its multilingual element. As noted earlier in this book, previous studies examining tourism translation often focus on a single language pair. The present corpus and the two case studies that will be based on it, will consist of more than one language combination, many of which are underexamined, and is the first multilingual corpus of its kind.

The texts collected consist of non-translated texts in American English, British English, French, Greek, and Russia and translations of these into the other languages (i.e. French into English, Greek into Russian, etc.) where available. Regarding English, a strategic decision has been made to include websites promoting the UK and the USA. These two particular varieties of English have been selected because they represent very popular English-speaking destinations, while at the same time people living in the UK and the USA engage in a lot of outbound tourism at least towards countries such as France and Greece, which are also included in the corpus (e.g. compared to Australia or New Zealand for example). This also allows for comparisons to be made between these two varieties to examine whether or not there is evidence of a rather uniform English tourism discourse. Regarding translations, although translated versions of websites often offer the option to choose between different varieties of English, in reality, with the exception of practical information (such as how to reach the destination), there is often no other difference in the language used, thereby reflecting a so-called international version of English, whatever this might mean linguistically. It might be the case that English tourism discourse is indeed rather uniform, something which will be examined in the two case studies in this book.

The discussion of the corpus-building process needs to begin with the examination of the available pool of data, that is the available official tourism websites. Due to their online nature, websites are time-sensitive and are continually updated, replaced, or even in extreme cases discontinued. As a result, any material collected from them is only a snapshot of tourism discourse at that specific point in time, although there is little reason to believe that this discourse will dramatically change over a short period of time. Significant changes in non-translated text might, however, occur, but they are likely to take time, as stereotypes in tourism discourse are difficult to shift and may result in frustration in reader-tourists (Travlou 2002). Overall, although the data collected as part of the TrAIL corpus capture a specific point in time, it is possible to argue that any patterns observed in the data are representative of that time period (e.g. not just that particular year but a few years before and after data have been collected), while remaining cautious regarding generalisations outside that general time frame.

The data comprising the TrAIL corpus represent the post-pandemic discourse of tourism and were all collected in the second half of 2021. Apart from the obvious fact that this was the most recent year that could realistically be captured to allow for in-depth analysis of data, there is also the important factor of the COVID-19 pandemic. Because during this time tourism came to a complete and unprecedented standstill, it gave the tourism industry the opportunity to pause and reflect on its practices and to consider better ways to attract visitors once travelling resumed. While this did not necessarily result in drastic changes which would improve the industry as a whole, it certainly gave destination promoters the opportunity to go through their websites and a number of instances have been noted where official tourism websites have been updated during that period. Further evidence of how current affairs might impact the availability of tourism websites comes from the poor availability of Russian texts. Identifying translations into Russian (especially from English and French) has generally been a challenge during the corpus collection process, but it would have been even more problematic had these texts been collected after March 2022 due to the Russo-Ukrainian war. After that time, some English websites (e.g. *Visit Britain*) ceased to publish their Russian version, while some Russian websites (e.g. *Russia Travel*) withdrew all their international versions. This shows that tourism texts are part of the fabric of society influenced by what is happening in it, which on a practical level also has significant implications for the data collection process.

Even without world events affecting the availability of data, the data collection process was not straightforward, especially for a multilingual corpus of this kind, where balance (to the extent that this is possible) is a priority. In many cases, the translated version of the websites consists of fewer pages or it might not be available at all. This also reflects how little attention is paid to translated versions of official tourism websites, and as a result to tourism translation more generally. When examining which websites to include in the corpus, the following criteria were followed: (1) that it is an official tourism website produced by either a public or private organisation presenting either the entire country or a specific city or region, (2) that translations in as many of the other languages examined here (i.e. English, French, Greek, Russian) are available where possible, and (3) that these translations have not been produced exclusively by machine translation based on the information available on the website. The starting point was always the official country website, before moving on to regions or cities, depending on the particularities of each locale. An additional practical consideration was for there to be enough available data, at least for the non-translated component, which meant prioritising cities or regions that are the most popular.

While these criteria might appear straightforward, the latter two relating to translation proved to be particularly problematic, especially regarding official English-language tourism websites. As an example, British English official tourism websites are discussed in detail here to illustrate how translations are often scarce or produced by machine translation. While the official tourism website for the UK existed at the time of collection in multiple language versions, including French and Russian, looking into the individual nations, only the official website for Scotland was available in other languages, including French, and the equivalent website for Wales was only available in Welsh and German, apart from English, and therefore could not be included. The official tourism website for England was only available in English, while the one for Northern Ireland provided translated versions 'powered by Google Translate'. Next, official tourism websites for counties and wider regions were examined and in total 69 were identified. Of these, 43, that is approximately 62%, were only available in English, while 18 (26%) offered translations in other languages 'powered by Google Translate'. This included websites for popular regions such as the West Midlands and Yorkshire. Only 8 websites (12%) offered another language version which appeared not to have been produced exclusively by machine translation. Of these, an even smaller number included a French or a Russian version or even both. Therefore, although a good amount of data appeared to be available at the beginning of the corpus-building process, in the end only a very small number of websites could be considered. This demonstrates beyond any doubt how little attention is paid to tourism translation by official tourism bodies in the UK and how little money they are prepared to invest in it, while at the same time the design of some of these websites suggests a significant investment. In the end, the three official websites selected for British English were *Visit Britain*, *Visit Scotland*, and *Visit London*. A similar picture was obtained when examining American English websites originating in the USA: out of the 71 websites identified, 36 (51%) were only available in English, 13 (18%) offered a translated version by Google, while 22 (31%) also included another language version, although French and Russian were not the typical options, with Spanish being by far the most popular. While American websites seem to fare a bit a better than British ones in providing international versions of their official tourism websites, the same problem persists: translation is the exception, rather than the norm.

As might be expected, the situation improves when examining French, Greek, and Russian non-translated texts, at least regarding the existence of an English version. However, things become more complex when examining whether a version in any of the other two languages also exists. Specifically, no Greek translations were identified and it was particularly difficult to identify French translations of Russian websites, with none being included in the end due to their scarcity, while a limited number of Russian translations of French websites has been included.

4.1.2 Corpus Design

Based on the discussion above about the focus of the corpus and the availability of the data, the resulting TrAIL corpus consists of 15 components: non-translated tourism websites (5 languages, hence 5 components), translations of the websites in each language/language variety into French (3 components), Russian (4 components), and English (3 components). These can be considered the building blocks of the corpus and are presented in Table 4.1. Depending on how these components are combined, they can create different subcorpora, which are presented in detail later in this section. At this point, it is worth noting that the non-translated texts are also the source texts of the translations. The corpus is relatively large, consisting of approximately 2 million words, which has been argued to be a suitable size for a specialist corpus (Nelson 2010,

56). This is one of the largest ever compiled to date for the study of tourism translation and includes the largest parallel element. A unique aspect of this study, which also explains the large corpus size, is that it is not limited to a specific language pair, but rather covers at least six different language combinations between Greek, English, French, and Russian.

Some components, however, are smaller than others due to the lack of availability of translated data. In some cases, as for example with Russian translations from British English, these were all the texts that could be collected during the time, that is they constituted the entire Russian version of the website, instead of a representative sample as with other websites. A larger corpus size, particularly regarding the non-translated components, where data collection has been easier, would have resulted in an even greater imbalance between translated and non-translated

Table 4.1 TrAIL corpus components

Corpus component	Abbreviation	Number of words
1. British English non-translated texts	EN_UK NT	170,704
2. American English non-translated texts	EN_US NT	170,520
3. French non-translated texts	FR NT	170,06
4. Greek non-translated texts	EL NT	170,578
5. Russian non-translated texts	RU NT	169,500
6. French translations of the British English texts	EN_UK-FR	128,742
7. Russian translations of the British English texts	EN_UK-RU	43,707
8. French translations of the American English texts	EN_US-FR	196,771
9. Russian translations of the American English texts	EN_US-RU	53,430
10. English translations of the French texts	FR-EN	149,411
11. Russian translations of the French texts	FR-RU	36,464
12. English translations of the Greek texts	EL-EN	170,396
13. French translations of the Greek texts	EL-FR	130,623
14. Russian translations of the Greek texts	EL-RU	88,461
15. English translations of the Russian texts	RU-EN	186,169
Total		2,035,536

texts. On the other hand, although allowing the availability of translated data to guide corpus size would have resulted in a more balanced corpus, it would have been a very small one. Therefore, a compromise has been reached as a result of practical considerations, which are very common in corpus studies, and the guiding principle was for the size of non-translated components to be around 170,000 words each.

To achieve diversity in each corpus component, texts from three official tourism websites were included in each non-translated category, following the criteria described in the previous section. Although diversity is not particularly hard to achieve regarding non-translated texts, it is significantly harder, as already discussed, with translated data and that is why some corpus components are much smaller (e.g. Russian translations from French consist of only one website), while others (e.g. French translations into Greek) are non-existent. This kind of imbalance is typical in corpus studies, especially multilingual ones, where there is often "a compromise between the hoped for and the achievable" (ibid., 60). This has an effect on representativeness and the reliability of the quantitative findings, which will therefore be taken into account when discussing results. However, this inconsistency does not affect the comparability of corpus data, as both raw and normalised frequencies are reported. Table 4.2 offers a detailed list of the websites included in the corpus. Weblinks are provided for each website, which was functional at the time of data collection.

It is worth remembering here that the starting point is always the main official tourism website for the country before moving on to specific cities or regions, the most popular of which are prioritised, as they are likely to include more data and be translated into more languages, as previously noted. It is interesting that the only case where there exist two official tourism websites for the country is Greece. While *Visit Greece* is the website created by the Greek National Tourism Organisation, a public organisation linked to the Ministry of Tourism, *Discover Greece* is created by Marketing Greece, a collaboration of the private tourism sector in Greece, a non-profit initiative of the Greek Tourism Confederation and the Hellenic Chamber of Hotels. This is the only instance where a website produced by a private body has been included in the TrAIL corpus. Its aims, however, appear to be the same as those of websites

Table 4.2 Official tourism websites in the TrAIL corpus also showing languages available in translation

British English	Visit Britain	https://www.visitbritain.com	English	57,151
			French	37,884
			Russian	43,707
	Visit Scotland	https://www.visitscotland.com	English	56,506
			French	59,920
	Visit London	https://www.visitlondon.com	English	57,047
			French	30,938
American English	Visit USA	https://www.visittheusa.com	English	56,270
			French	71,902
	Miami & Beaches	https://www.miamiandbeaches.com	English	58,738
			French	68,198
			Russian	53,430
	Discover Los Angeles	https://www.discoverlosangeles.com	English	55,512
			French	56,671
French	Explore France	https://www.france.fr/fr	French	56,142
			English	43,667
			Russian	36,464
	Paris	https://parisjetaime.com	French	56,621
			English	54,201
	Provence-Alpes-Côte d'Azur	https://provence-alpes-cotedazur.com	French	57,297
			English	51,543
Greek	Discover Greece	https://www.discovergreece.com/el	Greek	58,287
			English	61,106
			French	69,247
			Russian	50,141
	Visit Greece	https://www.visitgreece.gr/el	Greek	56,047
			English	59,094
	Incredible Crete	https://www.incrediblecrete.gr	Greek	56,244
			English	50,196
			French	61,376
			Russian	38,320

(continued)

Table 4.2 (continued)

Russian	Russia Travel	https://russia.travel	Russian English	57,088 57,762
	Discover Moscow	https://discovermoscow.com	Russian English	57,403 72,506
	Visit Petersburg	https://www.visit-petersburg.ru	Russian English	55,009 55,901
Total				2,035,536

produced by public bodies, i.e. "to promote Greece as a modern and enticing tourism destination" (*Discover Greece*, About Us section). In other words, it does not promote the interests of the private sector (e.g. a specific accommodation venue) and it is not directly linked to financial gain for that sector, although of course an indirect benefit exists. The representation of Greece as a tourism destination on these two websites circulates simultaneously and is equally available to reader-tourists both within and outside Greece.

The 15 corpus components presented earlier were then combined in different ways to create a number of subcorpora to allow for a detailed examination of the language of tourism across non-translated texts, between translated and non-translated texts in the same language, and between source and target texts. The corpus structure is presented in Table 4.3. As noted earlier, the compilation process has been guided by a corpus data triangulation approach to facilitate all of the above comparisons. Triangulation is achieved by mainly combining different corpus types (comparable and parallel), but also different types of texts (translated and non-translated) and different languages (British and American English, French, Greek, and Russian). This allows for more reliable results to be reached which are not based on a single language pair, and for similarities and differences across languages to be captured. In total, the 15 different corpus components will be examined in various combinations as part of the two case studies. There is naturally some overlap and two subcorpora might share the same component. This is evidence of corpus data triangulation and of an integrated approach towards the analysis of the data.

Table 4.3 TrAIL corpus structure

Subcorpus	Components
A. Comparable, multilingual, synchronic subcorpus of non-translated tourism texts	• British English non-translated texts • American English non-translated texts • French non-translated texts • Greek non-translated texts • Russian non-translated texts
B. Comparable, monolingual (English) synchronic subcorpus of translated and non-translated tourism texts	• British English non-translated texts • American English non-translated texts • French texts translated into English • Greek texts translated into English • Russian texts translated into English
C. Comparable, monolingual (French) synchronic subcorpus of translated and non-translated tourism texts	• French non-translated texts • British English texts translated into French • American English texts translated into French • Greek texts translated into French
D. Comparable, monolingual (Russian) synchronic subcorpus of translated and non-translated tourism texts	• Russian non-translated texts • British English texts translated into Russian • American English texts translated into Russian • French texts translated into Russian • Greek texts translated into Russian
E. Parallel, bilingual (French-English), synchronic subcorpus of tourism texts	• French non-translated texts • English translations of the French texts
F. Parallel, bilingual (Greek-English), synchronic subcorpus of tourism texts	• Greek non-translated texts • English translations of the Greek texts
G. Parallel, bilingual (Russian-English), synchronic subcorpus of tourism texts	• Russian non-translated texts • English translations of the Russian texts

(continued)

Table 4.3 (continued)

Subcorpus	Components
H. Parallel, bilingual (American English-French), synchronic subcorpus of tourism texts	• American English non-translated texts • French translations of the American English texts
I. Parallel, bilingual (American English-Russian), synchronic subcorpus of tourism texts	• American English non-translated texts • Russian translations of the American English texts
J. Parallel, bilingual (British English-French), synchronic subcorpus of tourism texts	• British English non-translated texts • French translations of the British English texts
K. Parallel, bilingual (British English-Russian), synchronic subcorpus of tourism texts	• British English non-translated texts • Russian translations of the British English texts
L. Parallel, bilingual (French-Russian), synchronic subcorpus of tourism texts	• French non-translated texts • French translations of the French texts
M. Parallel, bilingual (Greek-Russian), synchronic subcorpus of tourism texts	• Greek non-translated texts • Russian translations of the Greek texts

Corpus data triangulation as applied here is a combination of sequential and simultaneous analysis. It is sequential because analysis of the Subcorpus A (comparable, multilingual, synchronic subcorpus of non-translated tourism texts) will reveal between which languages differences are evident and this will further guide which of the remaining subcorpora need to be examined. It is simultaneous because once cross-linguistic differences are established, the remaining subcorpora will be examined in parallel—the results obtained from each one providing the answers to the research questions outlined below. Overall, non-translated tourism texts are compared to each other, followed by a comparison between translated and non-translated texts in the same language, and finally a comparison between source and target texts.

4.2 Methods of Analysis

Most websites included in the TrAIL corpus consist of several subsections, the lengthiest of which are *Places to Go* and *Things to Do*, which are typical of tourism websites. There are, however, some notable exceptions. Specifically, none of the Russian official tourism websites examined has a *Things to Do* section which might reflect how little emphasis is generally placed on tourist behaviour compared to descriptions of the destination. However, they do focus on promoting the physical space, with sections dedicated to places to visit. Similarly, some English-language official tourism websites, namely *Visit London*, *Miami and Beaches*, and *Discover LA*, focus specifically only on things to do, rather than places. This might be explained by the fact that a specific region is promoted instead of an entire country, which offers limited options for references to specific locations or because within smaller areas activities become conflated with specific locations. A similar pattern for the same underlying reason is observed in *Incredible Crete*. Interestingly, the section focusing on what to do on the *Visit the USA* official tourism website is titled *Experiences*, reflecting the importance of not just action but also how a destination is perceived through all the senses, a key component of tourism discourse, as the first case study will show. Of course, references to things to do can be found in sections introducing a specific place to the reader-tourist and vice versa, descriptions of places can be found in sections about activities. What is important here is the recognition that these two components, that is, tourist behaviour and destination description, are two of the most prominent ones in promotional tourism discourse.

Consequently, the present book focuses on two separate case studies. Case study I (Chapters 5–7) examines how tourist behaviour is conceptualised across societies which are experiencing different phases of tourism. The focal point of this case study is the verbal group and specifically verbs where the reader-tourist is involved as the explicit or implied subject/agent. Below are a few examples of sentences that will be analysed as part of Case study I. Items in bold indicate the verb that is the focus in each case.

After a short boat crossing from Sandwick, you can ***climb*** *the narrow steps to the top of the broch and take in breathtaking views across Mousa Sound* (Source: www.visitscotland.com)
Looking for *a quite [sic] spot among the West End's hustle and bustle?* (Source: www.visitlondon.com)
Make sure you ***sample*** *some of Plymouth's famous gin at the Black Friars Distillery.* (source: www.visitbritain.com)

Case study II (Chapters 8–10) examines how physical space is framed differently across languages and what that means for the way different people categorise and gaze at the world around them. The focal point here is the adjectival group and specifically adjectives that modify nouns referring to elements of physical space. Below are a few examples of sentences that will be analysed as part of case study II. Items in bold indicate the adjective that is the focus in each case.

Below are some of London's ***biggest*** *and* ***best*** *parks.* (Source: www.visitl ondon.com)
From acres of flowers to ***historic*** *Japanese gardens and* ***lush*** *hidden gems, read on and discover the* ***best*** *gardens in L.A.* (Source: www.discoverlosa ngeles.com)
Portmeirion, the ***unique*** *Italianate village created in the late 20th century, is a must-visit with its* ***picturesque*** *buildings and* ***beautiful*** *sub-tropical gardens.* (Source: www.visitbritain.com)

Details of the methodology employed for each case study will be provided in the relevant chapters. In this section, some general methodological considerations will be discussed which are pertinent to both explorations.

Although different in the nature of the linguistic features they examine (i.e. verbs and adjectives, respectively), the two case studies attempt to answer the same set of overarching research questions, which are outlined below:

1. Are linguistic features distributed differently in non-translated tourism texts in American English, British English, French, Greek, and Russian?

2. Where differences are observed, what types of adaptation take place during translation?
   a. How do the target texts compare to non-translated texts in the same languages?
   b. What role do source texts play?

The software selected for the first stage of the analysis of the corpus data is Sketch Engine (Kilgarriff et al. 2014), because it allows for Part-of-Speech (POS) tagging of different corpus components. The Wordlist tool is used to generate a list of all relevant linguistic items (i.e. verbs for the first case study and adjectives for the second one), which are then sorted based on their frequency. Each entry in the frequency list is then examined separately: a concordance list for each verb/adjective is generated by selecting it from the frequency list. Each entry in the concordance list is examined separately and in detail to establish the subject/agent for each clause when verbs are investigated (case study I) and the noun when adjectives are investigated (case study II). Concordance lines where the reader-tourist is not the subject/agent were excluded from analysis, as were those were the noun modified by the adjective is not related to physical space.

Because of the large number of both verbs and adjectives in the corpus, and also because, as explained above, each concordance line needs to be analysed separately to establish the context only the most frequent verbs and adjectives have been included in the analysis. Specifically, the analysis detailed above, that is, the examination of concordance lines for each verb/adjective, continued until a cut-off point of 18 words per 100,000 (0.18 per thousand words [ptw]) was reached, meaning that for a corpus segment of, say, 170,520 words, some 30 adjective or verb types would be selected. This number has been chosen for both research and practical reasons. On the one hand, repetition of relevant verbs/adjectives is likely to reflect key notions of tourism, while it is also an important element of promotion. Therefore, by focusing on more repetitive items, it is possible to better understand the most important elements of that particular tourism text. On the other hand, a cut-off point is needed to make corpus investigations manageable, especially given the size of the TrAIL corpus. For all these reasons, a cut-off point of 0.18 ptw was considered adequate. Based on trial examinations of the different corpus components, this was considered a good cut-off point from a practical point of view, allowing for enough types to be captured—this corresponds to approximately 30 verb types in the largest corpus components and to 10 in the smallest ones. This approach allows results to be comparable across corpus components. Once data are obtained, their normalised frequencies, expressed as percentages, are calculated, and detailed tables are created. These are so that readers get a sense of relative frequencies and do not play any role in the statistical tests conducted, which are explained later.

As will be explained in subsequent chapters, for each linguistic category examined (i.e. verbs and adjectives) a break down into semantic categories such as Motion, Discovery for verbs, and Time and Size for adjectives, will be provided, the most prominent of which will be examined in more detail. For each of these categories, it will be possible to establish the type of adaptation observed in the translated texts. In some cases, it will also be possible to make a case for an overall adaptation pattern for that particular language pair if a consistent approach is observed. In sum, the methodology employed for the two case studies consists of a calculation of frequencies, their statistical interrogation, and finally, a more in-depth qualitative examination of specific patterns where relevant. This applies to both the examination of cross-linguistic differences in non-translated texts and also the examination of adaptation patterns in translated texts.

As noted, corpus method triangulation is used for the analysis of data, and specifically within-method sequential triangulation. This is achieved by combining two types of descriptive statistics (raw frequencies and normalised frequencies) and two types of inferential statistics (log-likelihood test and Bayes Factor). Descriptive statistics, which are generated first, aim at providing an overview of the distribution patterns: they simply report on the frequencies of the categories. Inferential statistics, on the other hand, aim at revealing whether any differences observed in the distribution patterns are due to the inherent variability in the corpus components or are attributable to some other factor. In simple terms, descriptive statistics describe, while inferential statistics help interpret the significance. Where relevant, a qualitative analysis is also conducted to identify specific patterns in the data, therefore there is also an element of between-method triangulation. Details of how descriptive statistics have been calculated and what the ratio of occurrence is for each case study will be provided in the relevant chapters.

The inferential statistical tests are conducted on the raw frequencies and are generated after the descriptive statistics have been obtained; they involve a log-likelihood test and the Bayesian Information Criterion (BIC) as an approximation method for the Bayes Factor. These are conducted with the help of the log-likelihood and Effect Size Calculator (https://ucrel.lancs.ac.uk/llwizard.html), developed by the University of Lancaster. This is a user-friendly calculator with a simple 2 × 2 table, which is populated with relevant data, and allows for comparisons between two corpora (or corpus components as is the case here). In the first column, the frequency of the word under investigation is inputted (for example, the number of Motion verb tokens), as well as the size of the corpus component (for example, the total number of verb token identified). The same then applies to the second column for the other corpus component. The calculator generates data in the form of a table presenting a number of outputs, such as percentage difference, effect size, and odds ratio, but the two measures used for the analysis of data in the present case studies are: the log-likelihood test (*G*-test), which helps establish whether or not the differences observed in the data are the result of inherent variability in the corpus, and the Bayes Factor (BIC), which measures the probability that a difference in frequency is due to chance. The null hypothesis (H_0) is that any differences observed in the corpus components are due to chance, while the alternate hypothesis (H1) is that the difference found can be attributed to a factor other than chance (e.g. differences in linguistic conventions or influence from the source text, etc.) depending on the corpus components being examined. A 95% certainty is generally acceptable for linguistic analysis, that is, the *p* value (calculated as part of the log-likelihood test) needs to be less than 0.05. If that is the case, the observed difference is most probably not due to chance and therefore the null hypothesis can be refuted. The online calculator generates a log-likelihood (LL) score, which can then be associated with a certain *p* value. Detailed information on the LL scores and the associated *p* values is provided on the calculator main webpage, but suffice it to say here that a LL score of 3.84 or higher is needed for a *p* value of 0.05.

To strengthen further the arguments based on the corpus data, the BIC is also used. BIC is a criterion for model selection, but in the case of corpus-based studies can provide further evidence against or in favour of the null hypothesis, essentially circumventing any discussion about the differences in confidence levels among $p < 0.05$, $p < 0.01$, etc., since even with a *p* value of 0.05, there is still a 5% possibility that the results might be due to chance. BIC can take any value (both positive and negative). Positive figures offer evidence against the null hypothesis, while negative figures offer evidence in favour of it. The webpage for the online calculator developed by the University of Lancaster provides a scale for the interpretation of these values with corpus data, which is attributed to Wilson (2013). Accordingly, a figure between 0 and ± 2 is not worth more than a bare mention, and is therefore not considered as bearing any statistical significance, a figure between ± 2 and ± 6 is considered positive evidence, a figure between ± 6 and ± 10 is considered strong evidence, and any figure larger than ± 10 is considered very strong evidence against/in favour of the null hypothesis. The reason why both the log-likelihood test and BIC are needed is that the log-likelihood test can tell us whether or not a difference is due to chance, while BIC can tell us how sure we can be about it. Differences are considered to be statistically significant, i.e. two corpus components are regarded as 'different', if the *p* value is less than 0.05 and the BIC is higher than 2. No importance is attached to the degrees of variance in BIC values, that is, a BIC value of 11 is interpreted in the same way as one of 6—in both cases the null hypothesis can be refuted. Occasionally, when interpreting results, some judgements might be made regarding how similar (or not) distribution patterns are in different corpus components; for example, Motion verbs are used with a similar frequency in American English and British English non-translated tourism texts. Such comments are largely based on descriptive statistics, that is, normalised frequencies, and are the initial stage of comparison. More nuanced and robust judgements are made when results from inferential statistics are obtained.

The inferential statistical tests are conducted pair-wise and for each semantic category separately. For example, when examining the frequency distribution of Motion verbs in non-translated texts, the raw frequencies of these verbs are compared between American English and British English, American English and French, American English and Greek, and so on. Similarly, when examining, for example, the frequency distribution of Motion verbs in translated English texts, the raw frequencies of these verbs are compared between American English non-translated texts and English texts translated from French, between American English non-translated texts and English texts translated from Greek, and so on. Although a comparison across all non-translated texts is possible, this will only reveal whether or not Motion verbs, in this particular example, are distributed differently across languages, and not any similarities or differences between specific languages. Pair-wise comparisons also allow readers to focus on the specific language pair(s) they are interested in, which is bound to be useful for both trainee and practicing translators.

Overall, the two case studies make use of multiple triangulation by employing both corpus data and corpus methods triangulation. Corpus data triangulation establishes the different stages of the analysis and which subcorpora will be examined at each stage, while the comparison is achieved through corpus method triangulation. Below are some worked examples of how exactly quantitative data have been generated and interpreted in terms of statistical significance. They capture different possible scenarios, and are all taken from the comparison of non-translated tourism texts. An example from the examination of translated texts will be provided later, when adaptation patterns are discussed.

Example for significant result:
In the corpus component of American English non-translated tourism texts which consists of 170,520 tokens, 2,750 verbs which meet the threshold frequency measure of 0.18ptw and which also have the reader-tourist as their subject/agent have been identified. Out of these verbs, 646, that is 23.49%, belong to the semantic category of Motion. In the corpus component of Russian non-translated tourism texts, which consists of 170,881 tokens, 809 verbs have been identified, following the criteria defined above. Of these, 320, that is 39.56%, are Motion

verbs. The difference in the frequency of Motion verbs between the American English and Russian corpus components is therefore compared to establish whether it is statistically significant using the online calculator developed by the University of Lancaster. The data are compared as follows: 646/2,750 vs 320/809. The LL value is 54.34, suggesting that the probability p value is less than 0.05, while at the same time the BIC score is high at 46.17, providing very strong evidence against the null hypothesis. Considering these two inferential statistics together, we can conclude that Motion verbs are distributed more frequently in the Russian non-translated tourism texts than in the American ones and that the difference in their frequency is not due to chance but rather, for example, related to genre preferences.

Example for non-significant result:
We can now compare the above frequency of Motion verbs in American non-translated tourism texts to that noted in the Greek corpus component, which consists of 170,601 tokens. Here, 1,503 verbs have been identified which meet the same criteria (i.e. that of frequency and subject/agent), of which 394, that is 26.21%, encode Motion. When comparing these frequencies, that is 646/2,750 and 394/1,503, the log-likelihood value is 2.92, which is lower than the critical value of 3.84 for a probability value of 0.05, while the BIC score is negative at −5.44, providing positive evidence in favour of the null hypothesis. Therefore, it can be concluded here that the frequency of Motion verbs in the American English and the Greek tourism texts is not statistically significant and any observed difference is due to chance.

Other possibilities:
It is also possible for the log-likelihood test to provide evidence that the difference is significant, but for the BIC score not to provide sufficient

evidence against the null hypothesis. Since triangulation is adopted in this case, such cases are considered not to provide sufficient evidence to argue with a relative degree of certainty that the difference is not due to chance, and therefore are not considered statistically significant. One such example is the comparison of the frequency of Motion verbs in the American English and the French non-translated tourism texts. In the French corpus component, which consists of 170,060 tokens, 2,018 verbs have been identified according to the criteria defined earlier. Of these, 535 verbs, that is 26.51%, denote Motion. When these frequencies are compared, that is 646/2,750 and 535/2,018, the log-likelihood score is 4.26, which is higher than the critical value of 3.84 for a probability value of 0.05, but the BIC score is −4.21, providing positive evidence in favour of the null hypothesis. Since the null hypothesis cannot be, in this case, refuted, the difference in the frequency of Motion verbs in the American English and the French non-translated tourism texts is considered to be most likely a result of the inherent variability of the corpus components, rather than a distinguishing preference of tourism texts in any of those two languages.

Finally, since the focus of the explorations is to a large extent on adaptation in translation (Research Question 2), a refined framework has been created to understand in more depth how this can be demonstrated. Although it might be expected that there is a simple dichotomy between adaptation and no adaptation, in reality it is possible to identify different types of adaptation. The first thing to consider is whether adaptation is expected to be found, which is based on comparisons made between non-translated texts in the language pair in question. If non-translated texts exhibit similarities, then adaptation is not expected during translation. If, however, they exhibit differences, then adaptation is expected. It is also possible that unique types of adaptation are noted, such as ‘over-adaptation’ and ‘reverse adaptation’. To illustrate what these might entail, we can take a hypothetical example of Motion verbs being more frequently used in non-translated tourism texts in language A than language B. When tourism texts are translated from language A into language B, it is expected that Motion verbs will be less frequently found in target texts than source texts, following the communicative preferences of tourism texts in the target language. However, if Motion verbs are found even less frequently in the target texts than in non-translated texts in language B, then this is an example of over-adaptation or, in other words, of too much adaptation. If Motion verbs are found to be used more frequently in target texts than in source texts, then this is an example of reverse adaptation. Both over-adaptation and reverse adaptation are cases of unsuccessful adaptation.

Table 4.4 summarises the types of adaptation that can be anticipated and explains how they can be methodologically established.

An example of how inferential statistics are used to establish adaptation patterns is provided below, once again taken from the examination of Motion verbs.

Example for statistical significant result for adaptation:
Considering the discussion of the previous examples, it can be shown that the frequency of Motion verbs in American English and Greek is similar and therefore no adaptation is expected in this case (no significant difference was found). This means that when the Greek tourism texts are translated into English, the frequency with which Motion verbs are employed is expected to be similar between:

- Greek source (=original, non-translated) and English target texts (=translations), and by extension
- target texts (=translations) and non-translated tourism texts in American English.

In the corpus component of English translations from Greek, which consists of 170,369 tokens, 1,660 verbs have been identified which meet the required criteria. Of these, 597, that is 35.96%, are Motion verbs. This percentage is higher than that found in both American English and Greek non-translated tourism texts. The comparison between source and target texts (394/1,503 vs 597/1,660) reveals that the log-likelihood score is 24.15, suggesting that the probability p value is less than 0.05, while at the same time the BIC score is high at 16.01, providing very strong evidence against the null hypothesis. Therefore, the difference in the frequency of Motion verbs between the Greek source and the English target texts is statistically significant, and adaptation is considered to have

Table 4.4 Types of adaptation

Category	*Description*	*Quantitative measurements*
Not required	Adaptation is not expected and not observed. Similarities between non-translated texts in the two languages have been noted	• no statistically significant difference between non-translated texts in the two languages and at least positive evidence in favour of H_0 • no statistically significant difference between translations and non-translated texts in the target language and at least positive evidence in favour of H_0 • no statistically significant difference between the source and the target texts and at least positive evidence in favour of H_0
Additional	Adaptation is not expected but is observed. Similarities between non-translated texts in the two languages have been noted.	• no statistically significant difference between non-translated texts in the two languages and at least positive evidence in favour of H_0 • statistically significant difference between translations and non-translated texts in the target language and at least positive evidence against H_0 • statistically significant difference between the source and the target texts and at least positive evidence against H_0

(continued)

Table 4.4 (continued)

Category	*Description*	*Quantitative measurements*
No	No adaptation is observed but is expected. Target texts stay close to patterns found in source texts.	• statistically significant difference between non-translated texts in the two languages and at least positive evidence against H_0 • statistically significant difference between translations and non-translated texts in the target language and atleast positive evidence against H_0 • no statistically significant difference between the source and the target texts and at least positive evidence in favour of H_0
Successful	Adaptation is expected and is observed. Target texts come close to non-translated texts in the same language and move away from source texts.	• statistically significant difference between non-translated texts in the two languages and at least positive evidence against H_0 • no statistically significant difference between translations and non-translated texts in the target language and at least positive evidence in favour of H_0 • statistically significant difference between the source and the target texts and at least positive evidence against H_0

Category	*Description*	*Quantitative measurements*
Weak	Adaptation is expected and is observed to some degree. Target text stands between non-translated texts in target language and source texts.	• statistically significant difference between non-translated texts in the two languages and at least positive evidence against H_0 • normalised frequencies found in translated text are between those found in source texts and non-translated texts • statistically significant difference between translations and non-translated texts in the target language and at least positive evidence against H_0 • statistically significant difference between the source and the target texts and at least positive evidence against H_0

(continued)

Table 4.4 (continued)

Category	*Description*	*Quantitative measurements*
Over	Adaptation is expected but is observed to a larger degree. Target texts move away from both non-translated texts in target language and source texts. The direction of adaptation is towards non-translated texts in target language and beyond.	• statistically significant difference between non-translated texts in the two languages and at least positive evidence against H_0 • normalised frequencies found in translated texts are higher/lower than those found in BOTH source texts and non-translated texts, but closer to non-translated texts • statistically significant difference between translations and non-translated texts in the target language and at least positive evidence against H_0 • statistically significant difference between the source and the target texts and at least positive evidence against H_0
Reverse	Adaptation is required and is observed. Target texts move away from both non-translated text in target language and source texts. The direction of adaptation is in the opposite direction from the non-translated texts in target language.	• statistically significant difference between non-translated texts in the two languages and at least positive evidence against H_0 • normalised frequencies found in translated texts are higher/lower than those found in BOTH source texts and non-translated texts, but closer to source texts • statistically significant difference between translations and non-translated texts in the target language and at least positive evidence against H_0 • statistically significant difference between the source and the target texts and at least positive evidence against H_0

occurred here, even though it is not expected. To further confirm this, the English target texts are compared to the American non-translated tourism texts, where differences are found to be once again statistically significant (log-likelihood score is 55.55 and BIC score is 47.16). This is a case of observed adaptation even if it might not be expected based on the data obtained from the non-translated corpus components. In simple terms, the English translations of Greek source texts use Motion verbs more frequently than both their source texts and non-translated texts in American English.

The triangulation methodology that has been adopted in the case studies deals with naturally occurring data which would not have been obtained if only a single type of corpus (e.g. parallel corpus) or single method were to be used. This increases confidence in the results and the conclusions reached based on these findings. This does not mean that results should not be treated with caution where relevant, but rather that the data obtained are rich enough to allow for a more profound interpretation of the data, especially where contradictions might be observed. Because the phenomena under investigation are particularly complex, a corpus triangulation approach can offer a much more detailed and balanced picture, especially if multilingual data are examined as is the case here. More detailed discussion about the specific method employed for each exploration will be provided in the relevant chapters.

References

Kilgarriff, Adam, Vít Baisa, Jan Bušta, Miloš Jakubíček, Vojtěch Kovář, Jan Michelfeit, Pavel Rychlý, and Vít Suchomel. 2014. "The Sketch Engine: Ten Years On." *Lexicography* 1: 7–36.

Lee, C. 2016. "A Corpus-Based Investigation of Theme Choice in English Translations of Korean Online Tourist Texts—With Focus on Interactional Themes." *Perspectives: Studies in Translatology* 24 (2): 294–318.

Malamatidou, Sofia. 2018. *Corpus Triangulation: Combining Data and Methods in Corpus-Based Translation Studies*. London & New York: Routledge.

Nelson, Mike. 2010. "Building a Written Corpus: What Are the Basics?" In *The Routledge Handbook of Corpus Linguistics*, edited by Anne O'Keefe and Michael McCarthy, 53–65. Abingdon and New York: Routledge.

Travlou, P. 2002. "Go Athens: A Journey to the Centre of the City." In *Tourism: Between Place and Performance*, edited by S. Coleman and M. Crang, 108–27. Oxford: Berghahn Books.

Wilson, A. 2013. "Embracing Bayes Factor for Key Item Analysis in Corpus Linguistics." In *New Approaches to the Study of Linguistic Variability: Language Competence and Language Awareness in Europe*, edited by M. Bieswagner and Ko.S. Amei, 3–11, Frankfurt: Peter Lang.

Part I

Case Study I: Things to Do

5

Case Study I: Key Concepts and Approach

This case study examines the first of the two main components of tourism texts, i.e. things to do at a destination. The focus is more on *do* rather than *things*: what is examined here is not the activities as such (i.e. snorkelling, cycling, shopping, etc.) but *how* tourists can engage and interact with the destination. In other words, the focus of this case study is on how tourist behaviour, and by extension tourism more generally, is conceptualised across different languages, and how any observed differences between tourism texts in different languages are negotiated during translation. The focal point of this case study is the verbal group and specifically verbs where the reader-tourist is involved as the explicit or implied subject/agent. Examples of sentences that might be examined include *The visitor strolls along the canal* (where *the visitor* is the explicit subject), *Stroll along the canal* (where *you* is the implied subject), and *The museum can be visited by the public* (where *public* is the explicit agent, or if omitted, it is considered to be implied).

The main assumption in the present exploration is that tourism texts encourage readers to visit the destination through the frequent use of key verbs referring to how tourists engage and interact with a destination. Importantly, tourism texts in different languages are expected to

S. Malamatidou, *Translating Tourism*, Palgrave Studies in Translating and Interpreting, https://doi.org/10.1007/978-3-031-49349-2_5

encode tourist behaviour as it is conceptualised in the society in which they have been produced and therefore cross-linguistic differences are to be expected, raising important issues for their translation. Simply put, readers with different linguistic and cultural backgrounds are likely to interpret the world around them differently and have a different understanding of their own position within that world. As a result, they are likely to conceptualise tourism and what they can do at a destination differently. Indeed, a recent study (Hickey et al. 2022) conducted by Visit Britain on the motivations, influences, decisions, and sustainability of inbound travellers revealed that, for example, tourists from the Americas do not prioritise rest and relaxation, but destinations with a lot on offer and a wide range of experiences in one place, while tourists from Western Europe and North-East Asia place a lot of emphasis on food and drink options. These preferences are expected to be encoded in the linguistic conventions of non-translated tourism texts, while they should also be reflected in the translated versions to maximise the promotional potential of tourism texts. For instance, if reader-tourists from a specific linguistic and cultural group are found to value and prioritise relaxation (based on the examination of the non-translated tourism texts in that language), then the translated texts in that same language will need to be adapted to also focus on relaxation, if they want to attract that particular group, even though relaxation might not be highlighted in the corresponding source text.

5.1 Tourism Phase

While identifying how certain linguistic features might be used differently in tourism texts across languages is important, simply listing features is not enough. What is needed is a framework for their interpretation to help answer the question of why certain differences might be observed. For the purposes of this case study, the notion of the tourism phase is used as an analytical tool for the interpretation of corpus findings. Tourism phase refers to how tourism as a social activity evolves historically with the transformation of societies and references to it can be found in the work of Sulaiman and Wilson (2019), even though they do not elaborate on how exactly this notion is conceptualised. Some references to a shift in the tourism phase can also be found in O'Dell (2005) and Francesconi (2007). Based on the above, an attempt will be made to theorise this notion and explain how it will be used for the analysis of the TrAIL corpus data.

The principal idea is that the way tourism is perceived is not a static but rather an evolving notion. As more and more people engage with tourism activities, they start prioritising different elements and seek out different things, and their concept of tourism changes. It is important to note that the concept of tourism phase is related to the individuals engaging with tourist activities, and not the destination or locale, while any shift is a collective, rather than an individual one. Progression is linear and all societies who engage with tourism activities start from the same point and, in principle at least, they all have the potential of reaching the next phase. The starting point is what has been called the sightseeing, gazing, or mainstream phase of tourism (Sulaiman and Wilson 2019): looking at something new is bound to have an effect, and sightseeing is an important element of the tourism experience. This mainstream idea of tourism is reflected in the typical image of the tourist as someone with a camera around their neck and part of a large tourist group touring around in a bus. The second phase of tourism is the post-sightseeing, experiential, or alternative phase of tourism (Sulaiman and Wilson 2019). As people become more experienced in their travelling, they start looking down upon the sightseeing idea of tourism, which is often associated with "passivity, inauthenticity and artificial experiences (staged authenticity); superficiality; lack of respect for, and interest in, encountered placed and people" (ibid., 155). People start wanting more than simply looking at sights; on the contrary they "want to engage that world and *do things*" (O'Dell 2005, 27 emphasis in the original). Although no other phases have been identified so far, it is possible that more might emerge as societies and tourism practices continue to evolve.

From a cross-cultural perspective, it is important to clarify a few points. Firstly, whether or not a society reaches the post-sightseeing phase varies greatly and depends on a number of socio-political circumstances which affect the availability and popularity of mass tourism. Societies where mass tourism is a fairly recent phenomenon or where engagement with tourism is relatively low will still be experiencing the sightseeing phase, while societies where mass tourism is much more established and widespread will probably have already reached the post-sightseeing stage (Sulaiman and Wilson 2019). According to Francesconi (2007), Anglophone societies have already reached that stage. Secondly, tourism phases are a continuum, rather than a binary opposition, and transition from one phase to the other requires time. Therefore, it is possible that some societies demonstrate only some elements of the post-sightseeing phase or only to some degree. With time, it is likely they might also reach the next phase, but this is not guaranteed.

Thirdly, and related to both points above, it has been argued that certain parts of the world, e.g. the Western world (Francesconi 2007), might be collectively evolving through the phases of tourism. Indeed, many historical similarities can be observed with regard to tourism in Europe and North America. Specifically, modern mass tourism emerged around the same time, that is after the Second World War. Social, economic, and technological changes meant that tourism became accessible to many more people, and not just the upper and middle classes. Additionally, many societies saw an increase in paid holidays, as well as in wage rate, while the relative cost of travel fell, making travel affordable (Page and Connell 2014). This resulted not only in a significant increase in domestic tourism, but also in an expansion of international air travel. However, unique socio-political circumstances associated with different locales means that differences can be observed even within the Western world for example. This is why it is important to understand the basic profile of each society examined as part of this study: American, British, French, Greek, and Russian.

United States of America

Mass tourism in the USA greatly benefitted from the revolution of air travel—a key element in a country spanning almost 10 million km^2 and separated by sea from most other parts of the world. The rising middle-class as well as the establishment of paid holidays for many workers around that period also contributed to the emergence of mass tourism and had a significant impact on both domestic and outbound tourism. According to Dupont et al. (2012), the number of Americans going abroad rose by 700% from 1950 to 1970, at a rate of 10.8% per year. Early on, Americans were in a much better position (both from a financial and technological point of view) than many other parts of the world. Today however things are somewhat different. According to Boniface et al. (2021, 592), the USA is "an affluent but 'leisurely poor' society compared to most European countries", with the amount of leisure time reducing significantly from the 1970s until the present day. More working hours per week, and fewer days of paid leave compared to most European countries, means fewer opportunities for travel, with 30% of the population never going on holiday (Boniface et al. 2021, 592).

United Kingdom

After the end of World War II, social and economic changes increased the demand for both domestic and outbound tourism. A combination of increased affluence, holiday entitlement, and personal mobility (Boniface et al. 2021, 136–137) meant that more people could travel than ever before. The 1960s was a particularly good decade in that regard which evidenced the beginning of mass demand for holidays abroad. By the end of the 1970s Britons were spending more on holidays abroad than at home, and today up to 75% of the population takes a holiday (Boniface et al. 2021, 140).

France

Unlike the USA and the UK, where tourism has been significantly boosted by air travel, the main enabling factor of mass tourism in France has been the increased leisure time available to the French, as well as a number of social tourism initiatives, especially in the 1960s, enabling unemployed and poorer citizens to take a holiday (Boniface et al. 2021, 266). Another distinguishing factor of the French society is the wide distribution of residence secondaire (second homes) or a holiday home in the countryside or seaside which French people own. Boniface et al. (2021, 268) argue that such homes are an indication that most French still have strong rural roots. All this has resulted in a large demand for domestic tourism and only 10% of French travel abroad, a proportion lower than most European countries, while over 60% of the population spend their holidays in France.

Greece

At the time when mass tourism was introduced in the western world thanks to the more stable, if not improving, social and political conditions in the decades following World War II, Greece was under a dictatorship (1967–1974), while it was also coming out of a devastating Civil War (1943–1949). Like France, holiday homes are also very popular in Greece, while many Greeks also have very strong links with the rural communities in which their parents or themselves grew up, owning houses there and visiting them typically in the spring and summer. This has resulted in a very moderate engagement with tourism which continues to this day with approximately only half of the population engaging in tourism, and with only 10% of trips being abroad (Boniface et al. 2021, 346).

Russia

While mass tourism started becoming more popular and accessible in the western world, in the Soviet Union of the time domestic tourism was managed by the state-controlled trade unions mainly in the form of sanatoria and purpose-built holiday centres, while international trips were scarce as special permission was required. The collapse of the Soviet

> Union in 1991 had a significant economic impact on the population, meaning that people had less money to spend on leisure activities, but at the same time the amount of leisure time increased. Similar to France and Greece, a large number of Russians who live in cities have a holiday home or country cottage in one of the villages on the outskirts, which is what dominates domestic tourism. The subsequent economic recovery allowed more people than ever before to travel abroad but continuing economic and political instabilities mean that the situation is increasingly volatile.

The discussion of the socio-political circumstances for each society represented in the TrAIL corpus, although very brief, serves to illustrate two main points. Firstly, not all parts of the world have been introduced to mass tourism with the same ease or intensity. Secondly, even though all of the countries considered here, with the exception of Greece at the time of writing, are top tourist-generating countries (i.e. countries where tourists come from), there are significant differences in the amount of engagement with domestic and outbound tourism by the population. This means that the experience of tourism and how this has evolved historically is likely to differ significantly across the world, and even in the so-called Western world, differences might be observed. It also means that societies will find themselves at a different point on the tourism phase continuum. Although similarities might be observed, especially between societies belonging to the same geographical (Europe) or linguistic (Anglophone) space, generalising based on these can result in an over-simplified picture. That is why no assumptions will be made here about which tourism phase each society might be experiencing, and instead a data-driven approach will be adopted, where the linguistic data analysed will aim to reveal which phase is being experienced by each society.

Understanding and analysing the tourism phase currently experienced by each society and how this is manifested linguistically can provide a useful guide for adaptation during translation. Many previous studies have listed linguistic features that are distinctive of tourism discourse but might be used differently between languages. However, what most of these studies fail to do is to explain why such differences might exist apart from inter-linguistic structural differences or linguistic preferences. In the present case study, such differences are seen as meaningful and reflect how different societies conceptualise tourism and the type of tourism message they want to promote. As a result, translators should be able to more easily make necessary adaptations to the text, when it is clear what aim these changes need to achieve.

5.2 Methods of Analysis

Since the focus of this case study is the reader-tourist, and specifically their behaviour, the corpus analysis involves the close examination only of those verbs where the reader-tourist is the explicit or implied subject/agent. However, even with this focus in mind, the number of verbs within each corpus component is very large. This together with the fact that each concordance line needs to be analysed separately to establish the meaning of the verb (especially in the case of polysemous verbs) and its subject, means that it is not possible to examine all verbs in a corpus of this size. As explained in Chapter 4, only the top most frequent verbs, that satisfy the 0.18 ptw frequency criterion, have been included in the analysis. It is worth reminding the reader here how top verbs have been identified. Firstly, a lemmatised wordlist of verbs sorted by frequency was generated for each corpus component with the help of SketchEngine (Kilgarriff et al. 2014), which allows for uploaded corpora to be automatically part-of-speech (POS) tagged, significantly facilitating the identification of verbs. Following this, each verb was selected, which generated the concordance list for this verb. The concordance line for each clause was examined to establish whether the explicit or implied subject/agent corresponds to the reader-tourist. Concordance lines where the reader-tourist is not the subject/agent were excluded from analysis.

During the examination of concordance lines, each verb was analysed closely to identify potential instances of polysemy, since the focus of this exploration is on meaning rather than form. Although the corpus is specialised in so far as it focuses on tourism promotion, it is still possible that verbs are used with different meanings, each of which might be associated with a different element of tourism. For instance, *see* can mean both perceive with the eyes or understand. Similarly, *discover* can mean find or learn. Related to this point, verbs that participate in phrasal or delexical constructions were also examined, but as before I manually created a separate entry for each meaning (e.g. *take away*, *take in*, *take a look*). This process resulted in a list of verbs, with separate entries for distinct meanings, and their observed raw and normalised frequencies for each corpus component separately. It is worth stressing here that it is possible that a verb is included in the analysis based on only one of its senses.

Once all verbs and their frequencies had been identified after the cut-off point had been applied, each verb was assigned to a semantic category. These categories are loosely based on categories provided by Systemic Functional Linguistics (SFL) (Halliday and Matthiessen 2014). Once verbs are grouped in semantic categories, it is possible to identify which aspects of the tourist behaviour are highlighted more (or less) in each language/text type by examining which semantic categories are more (or less) prominent. This can then in turn provide information about the tourism phase that each society is currently experiencing. Table 5.1 presents the semantic categories identified in the TrAIL corpus and the verbs through which they have been found to be realised across all languages and types of texts (translated and non-translated). Once all verbs had been identified and assigned to semantic categories, the analysis focused on the overall distribution of the semantic categories and how they are realised through specific verb choices. Any comparisons made across corpus components were based on statistical analyses using the log-likelihood text and Bayes (BIC) factor (see Chapter 4 for more details on methodology). It is worth repeating here that whenever a statistical difference is reported when presenting the results in the next two chapters, this means that the probability value p has been found to be less than 0.05 and that the BIC factor has been found to be larger than 2. However, the detailed results of statistical tests will not be reported, to avoid repetition and number-heavy discussion, especially considering how many components need to be compared. In total, 23 comparisons have been made across the four semantic categories which were found to be most prominent (see Chapter 6): ten comparisons involving non-translated texts, nine comparisons involving texts translated into English, eight involving texts translated into French, and six involving texts translated into Russian. If each semantic category is counted as a separate comparison, there were altogether 92 comparisons. The full details of all statistical tests can be found in Appendix I.

A note is needed here to explain how the distribution of verbs in semantic categories is measured. Instead of comparing normalised frequencies for each semantic category out of all words in the corpus component, the number of identified verbs above the cut-off point has been applied, i.e. has been used as a basis for comparisons (e.g. Motion verbs represent 60% of all identified verbs). This is because normalised frequencies are likely to simply reflect the differences in the total number of verbs identified, which might be quite significant across languages. For example, a low normalised frequency of Motion verbs in Russian tourism texts might simply reflect the fact that very few verbs addressing the reader-tourist are used in the language, overlooking that in fact Motion verbs are one of the most prominent categories in Russian tourism discourse when compared to the total number of identified verbs. The focus here is not on number of occurrences, but how semantic categories relate to each other to conceptualise tourist behaviour, that is, on how the semantic categories are distributed in different components of the corpus.

Finally, the analysis is divided into two distinct stages, each corresponding to a separate chapter. The first stage of analysis involves the examination of non-translated texts in American English, British English, French, Greek, and Russian to establish how tourist behaviour, and by extension tourism more generally, is conceptualised in these languages, and also the tourism phase which is currently being experienced. The second stage of analysis consists of an examination of translated texts in English, French, and Russian out of several source languages available depending on the availability of data in the TrAIL corpus to examine how any observed differences between tourism texts in different languages are negotiated during translation and whether target texts promote an idea of tourist behaviour in accordance with the target or with the source language. It is assumed that successful communication

Table 5.1 Most frequent verbs with reader-tourist as subject/agent in all corpus components grouped into semantic categories by language

Category	English	French	Greek	Russian
Motion	climb, come, continue, follow, get, go, head, reach, stay, stop, stroll, tour travel, visit, walk, wander	aller (go), arpenter (survey), arrêter (stop), arriver (arrive), commencer (start), continuer (continue), diriger (head), embarquer (embark), emprunter (follow), faire (un tour) (tour), flâner (stroll), grimper (climb), marcher (walk), monter (climb), partir (leave), promener (walk), rejoindre (rejoin), parcourir (travel), passer (pass), prendre (take), rendre (head), séjourner (stay), suivre (follow), traverser (cross), venir (come), visiter (visit), voyager (travel)	επισκέπτομαι (visit), περπατάω (walk), φτάνω (arrive), ακολουθώ (follow), αναβαίνω (climb)	вернуться (return), взбираться (climb), добираться (reach), ехать (drive), заглядывать (drop in), направляться (head), начать (start), останавливаться (stop/stay), отправляться (go), побывать (visit), подниматься (climb), пойти (go), покататься (ride), попадать (get somewhere), посещать (visit), приезжать (arrive), приходить (come), прогуливаться (stroll), прокатываться (drive past), проходить (pass), прыгать (jump), путешествовать (travel), сходить (go)

(continued)

Table 5.1 (continued)

Category	English	French	Greek	Russian
Action	discover, do, explore, find, join, look (for), make (sure), miss, play, start, try	assister (attend), faire (do), essayer (try), laisser (allow), manquer (miss), participer (participate), rejoindre (join)	παραλείπω (miss)	воспользоваться (play), использовать (use), заниматься (do), поиграть (play), попробовать (try), присоединяться (join), принимать (take part), пропустить (miss)
Discovery	discover, explore, find, look (for)	chercher (search), découvrir (discover), explorer (explore), rencontrer (encounter), trouver (find)	ανακαλύπτω (discover), βρίσκω (find), συναντώ (encounter), γνωρίζω (meet), εξερευνώ (explore), αναζητώ (look for)	встречать (meet), искать (look for), исследовать (explore), найти (find), находить (find), познакомиться (meet), открыть (discover)
Possession	book, bring, buy, grab, pick, rent, spend (time)	acheter (buy), passer (du temps) (pass time), réserver (reserve)		арендовать (rent), взять (bring), забронировать (book), заказывать (order), купить (buy), принести (bring), проводить (spend time),
Physiology	eat, relax	détendre (relax)		отдыхать (rest), поплавать (swim), расслабляться (relax)

Category	English	French	Greek	Russian
Perception	admire, browse, catch, check, experience, feel, immerse, sample, see, spot, take (in), taste, try, watch	admirer (admire), apercevoir (notice), consulter (consult), déguster (taste), goûter (taste), observer (observe), plonger (immerse), regarder (watch), savourer (savour),vivre (live), voir (see)	βλέπω (see), θαυμάζω (admire), δοκιμάζω (taste), ζω (live)	взглядывать (look), испытывать (experience), любоваться (admire), наблюдать (observe), окунуться (immerse), осматривать (see), отведывать (taste), погружаться (immerse), послушать (lsten), посмотреть (look), полюбоваться (admire), понаблюдать (watch), посмотреть (watch), попробовать (taste), почувствовать (feel), проверить (check out), увидеть (see)
Cognition	discover, find, know, learn, see understand	apprendre (learn), découvrir (learn), initier (learn), savoir (know), oublier (forget)		забыть (forget), знать (know), изучать (learn), ознакомляться (get acquainted), подумать (think), познакомиться (get to know), узнавать (learn)

(continued)

Table 5.1 (continued)

Category	English	French	Greek	Russian
Desire	choose, need, plan want	choisir (choose), planifier (plan), souhaiter (wish), vouloir (want)	θέλω (want), επιλεγώ (choose)	выбирать (choose), захотеть (want), планировать (plan), предпочитать (prefer), спланировать (plan), хотеть (want)
Emotion	appreciate, enjoy, like, love, soak	aimer (love), apprécier (appreciate), (delve), profiter (enjoy), imprégner (soak)	απολαμβάνω (enjoy), αγαπάω (love)	любить (love), наслаждаться (enjoy), оценить (appreciate), побаловать (treat yourself)

and promotion of the tourism message is achieved when translated texts are adapted to how the target society conceptualises tourist behaviour, which corresponds to the tourism phase currently experienced by said society. Adaptation is expected between societies that have been found to experience different phases of tourism and therefore differ in their linguistic preferences. The further apart on the continuum of tourism phases societies are, the higher the degree of adaptation expected (see Chapter 4 for more details on the types of adaptation examined).

References

Boniface, B., R. Cooper, and C. Cooper. 2021. *Worldwide Destinations: The Geography of Travel and Tourism.* 8th Edition. London and New York: Routledge.

Dupont, B., A. Gandhi, and T. Weiss. 2012. "The Long-Term Rise in Overseas Travel by Americans, 1820–2000." *The Economic History Review* 65 (1): 144–67.

Francesconi, Sabrina. 2007. *English for Tourism Promotion: Italy in British Tourism Texts.* Milano: Hoepli.

Halliday, M. A. K., and M.I.M Matthiessen. 2014. *Halliday's Introduction to Functional Grammar*. London and New York: Routledge.

Hickey, K., N. Chan, and C. Warren. 2022. "MIDAS: A Global Report—Motivations, Influences, Decisions and Sustainability in a Post-Pandemic Era."

Kilgarriff, Adam, Vít Baisa, Jan Bušta, Miloš Jakubíček, Vojtěch Kovář, Jan Michelfeit, Pavel Rychlý, and Vít Suchomel. 2014. "The Sketch Engine: Ten Years On." *Lexicography* 1: 7–36.

O'Dell, T. 2005. "Experiencescapes: Blurring Borders and Testing Connections." In *Experiencescapes: Tourism, Culture and Economy*, edited by T. O'Dell and P. Billing, 11–33. Koge: Copenhagen Business School Press.

Page, S., and J. Connell. 2014. *Tourism: A Modern Synthesis.* London and New York: Routledge.

Sulaiman, M. Zain, and Rita Wilson. 2019. *Translation and Tourism.* Singapore: Springer.

6

Tourist Behaviour

The analysis begins with the examination of tourist behaviour in non-translated texts to uncover any cross-linguistic differences in the use of verbs where the reader-tourist is the explicit or implied subject/agent. The focus is on specific semantic categories of verbs which are most prominent in tourism discourse, their overall distribution, and how they are realised through specific verb choices. The examination of these categories will reveal a unique profile for each language examined here: American English, British English, French, Greek, and Russian. This is important for two reasons. Firstly, this profile will reflect how tourist behaviour, and tourism more generally, is conceptualised in each language, and secondly, and relatedly, it will reveal the tourism phase that each culture is currently experiencing. It is worth repeating here that the analysis is data-driven, that is the data will indicate which tourism phase is being experienced, rather than data-informed, as no initial assumptions are made, once the relevant data have been identified from the corpus. Results of this stage of analysis will inform the analysis of translated texts in the next chapter.

S. Malamatidou, *Translating Tourism*, Palgrave Studies in Translating and Interpreting, https://doi.org/10.1007/978-3-031-49349-2_6

6.1 Distribution of Verbs

The examination of the overall distribution of verbs which meet the threshold frequency measure of 0.18ptw and which also have the reader-tourist as their subject/agent, serves as a backdrop for the rest of the analysis: their total number will be used as a basis for calculating the frequency distribution of these verbs belonging to specific semantic categories. The overall distribution of such verbs across languages and what potential differences might mean will not be discussed in any detail here, that is, for example whether one language employs such verbs more frequently than another. This is not to say that such an analysis would not have been interesting; on the contrary, it can offer some insight into the use of the so-called ego-targeting technique (Dann 1996; Francesconi 2007; Sulaiman and Wilson 2019) whereby a high number of verbs where the reader-tourist is the subject create a stronger link between the reader and the destination aiming to make readers feel special. However, such an analysis needs to be combined with the examination of other linguistic strategies, such as direct form of address, informal tone, and imperative mood (Sulaiman and Wilson 2019), which would distract from the focus of the current exploration which is the examination of which tourism phase each culture belongs to and how that might be reflected in the tourism discourse.

Table 6.1 presents the distribution of the most frequent verbs (i.e. those that meet the threshold frequency measure of 0.18ptw) and which also have the reader-tourist as their subject/agent. For consistency with how the threshold frequency was calculated, normalised frequencies (nf) are also calculated per thousand words. It is worth noting that the table captures verbs across all the semantic categories identified in this case study; distinguishing between these will be applied at the next step of the analysis (Table 6.2). 'Tokens' refers to the total number of words in the relevant corpus component (i.e. the collection of non-translated tourism texts in each language, see Table 4.1). 'Verbs (types)' refers to the number of unique verbs identified, irrespective of how many times they might be used in total, and 'Verbs (tokens)' refers to the number of particular verbs identified.

Table 6.1 Normalised frequencies of most prominent verbs with reader-tourist as subject/agent in non-translated corpus components by language

	Tokens	*Verbs (types)*	*Verbs (tokens)*	
EN_US	170,520	35	*f*	*nf (ptw)*
			2,750	16.12
EN_UK	170,704	53	5,082	29.77
FR	170,060	27	2,018	11.86
EL	170,601	20	1,503	8.81
RU	170,881	14	809	4.73

As can be seen from Table 6.1, there is considerable variation across the languages examined here regarding how frequently non-translated tourism texts include mention of the reader-tourist. British English tourism texts show a strong preference towards such verbs, while Russian tourism texts employ them rarely. Tourism texts in the other languages sit somewhere in-between, with American English closer to British English and French and Greek tourism texts closer to Russian tourism texts. The application of the two statistical tests of log-likelihood and BIC (see Chapter 4.2) to compare the use of relevant verbs in paired corpus components in different languages, reveal that differences are significant for all pair-wise comparisons. It is evident that tourism texts in each language differ in terms of the emphasis they place on the interaction between tourist and destination with some languages showing a stronger preference to explicitly state how the tourist is to behave at a destination. The fact that such big differences are observed in the number of verbs found in each language component is the reason why in what follows the frequency distribution based on the total number of verbs will be reported for each semantic category, rather than the normalised frequency per thousand words. When the basis differs so markedly, normalised frequencies are likely to simply reflect the high or low use of the relevant verbs in general rather than the preference towards specific semantic categories.

Once the number of verbs has been established for each language component (i.e. the collection of non-translated tourism texts in each language, see Table 4.1), it is possible to calculate the frequency distributions across the various semantic categories. Nine categories have been identified, namely Motion, Discovery, Action, Physiology, Possession,

Perception, Emotion, Cognition, and Desire, with varying degrees of frequency; these are summarised in Table 6.2. The most prominent semantic categories are marked in bold.

Certain semantic categories, i.e. Motion, Discovery, Perception, and Emotion, represent a high normalised frequency. High frequency is understood here as anything above 10% in any one language. Anything below that number is considered low frequency which is the case for Action, Physiology, Possession, Desire, and Cognition. Due to their

Table 6.2 Normalised frequencies of semantic categories of most prominent verbs in non-translated corpus components by language as a percentage of identified verbs (tokens)

		EN_US	EN_UK	FR	EL	RU
Verbs (tokens)		**2,750**	**5,082**	**2,018**	**1,503**	**809**
Motion	Raw frequency	**646**	**1,011**	**535**	**394**	**320**
	Normalized frequency	**23.49%**	**19.89%**	**26.51%**	**26.21%**	**39.56%**
Discovery	Raw frequency	**600**	**1,095**	**515**	**474**	**59**
	Normalized frequency	**21.82%**	**21.55%**	**25.52%**	**31.54%**	**7.29%**
Action	Raw frequency	159	452	180	34	0
	Normalized frequency	5.78%	8.89%	8.92%	2.26%	0%
Physiology	Raw frequency	43	110	0	0	31
	Normalized frequency	1.56%	2.16%	0%	0%	3.83%
Possession	Raw frequency	152	241	0	0	41
	Normalized frequency	5.52%	4.74%	0%	0%	5.07%
Perception	Raw frequency	**525**	**1,116**	**393**	**347**	**270**
	Normalized frequency	**19.09%**	**21.96%**	**19.47%**	**23.09%**	**33.37%**
Emotion	Raw frequency	**272**	**454**	**245**	**182**	**0**
	Normalized frequency	**9.89%**	**8.93%**	**12.14%**	**12.11%**	**0%**
Desire	Raw frequency	252	271	45	72	42
	Normalized frequency	9.16%	5.33%	2.23%	4.79%	5.19%
Cognition	Raw frequency	101	332	105	0	46
	Normalized frequency	3.67%	6.53%	5.20%	0%	5.69%

prominence, the discussion that follows focuses only on the high-frequency semantic categories. This is not to say that low-frequency categories are not important, but rather that they play a secondary role in the conceptualisation of tourist behaviour—taken together they correspond to less than 25% of all verbs in the relevant corpus component in any one language. From a practical point of view, valid conclusions are also harder to reach when very low numbers are found. Figure 6.1 is a visual representation of the frequency distribution of the four most prominent semantic categories across languages.

In what follows, each of these four semantic categories will be examined in depth, both in terms of the normalised frequencies, but also, where relevant, in terms of how they are realised, i.e. which verbs are used to realise Motion, Discovery, Perception, and Emotion. This detailed examination will reveal whether and to what extent the non-translated tourism texts in the languages examined here conceptualise these key components of tourist behaviour differently and ultimately what phase of tourism each society is currently experiencing.

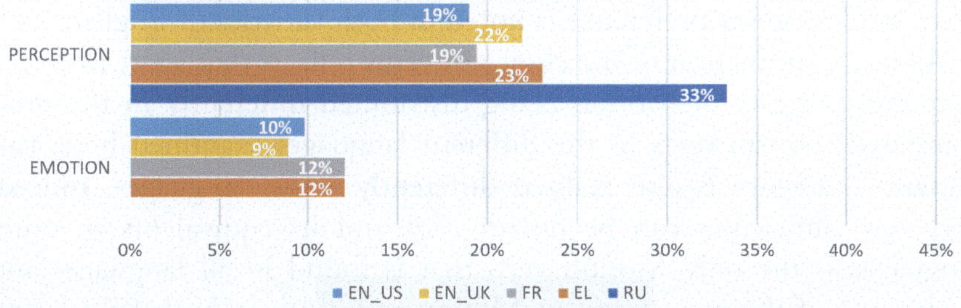

Fig. 6.1 Normalised frequencies of most prominent semantic categories in non-translated corpus components by language as % of all identified verbs

6.2 Motion

A large number of verbs identified denote Motion, which refers to the process of moving or changing position. This is not surprising considering that motion is a distinct characteristic of tourism as evident in its definition—"the phenomenon caused by the *departure* and the *return* of human beings from their place of habitual residence" (Panosso Netto 2009, 59 emphasis added). Being a prototypical characteristic of tourism means that a stronger emphasis on Motion reflects that a society is still experiencing the sightseeing phase, because the initial phase of tourism is more likely to be associated with its prototypical elements (e.g. taking photos and moving through locations).

Although Motion is a prominent category in all corpus components, some differences are observed in its distribution across languages, as Table 6.2 and Figure 6.1 show. Russian tourism texts show the strongest preference for this category (39.56%), followed by French (26.51%), Greek (26.21%), and American English texts (23.49%). In last place we find British English tourism discourse (19.89%). Statistical tests, conducted on the raw frequencies, reveal that differences in the distribution of Motion verbs across languages are significant in all cases where British tourism texts are compared with texts in any other language (expect for British vs American English), and also when Russian texts are compared with tourism texts in all other languages. This confirms that a very clear preference towards Motion is found in Russian tourism texts, while a lower preference is noted in both varieties of English.

A close examination of the choice of verbs reflects that apart from the semantic category of Motion being distributed differently in the non-translated tourism texts in the different languages examined here, this semantic category is also realised differently across languages. Indeed, very few similarities can be noted; *visit* and its equivalents in other languages is the only Motion verb that is found in all languages and is also one of the most frequent Motion verbs: 6th in both British and American English tourism texts, 5th in French, 1st in Greek, and 2nd in Russian. Visiting is an integral part of tourism, which is also reflected in the typical terms used to define those who engage with tourism (i.e. *tourist* but also *visitor*). While the similarities in the preference for *visit* across languages is striking and point towards the fact that there are cross-cultural similarities in the way tourist behaviour is conceptualised, the remaining Motion verbs suggest that overall Motion is perceived differently across languages.

Some languages conceptualise Motion in a leisurely relaxed way by frequently using verbs like *wander* in British English, *stroll* in American English, and *flâner* (*stroll*) in French, as illustrated in Example 6.1. The closest equivalent found in Greek tourism texts is περπατάω (*walk*), which is rather neutral in terms of meaning and does not capture the pleasure that is associated with strolling (see also section 6.5 on Emotion for a more detailed discussion of the role that pleasure might play in tourism discourse). It is interesting that equivalents of neither *stroll* nor *walk* have been found in the verbs frequently used in Russian tourism texts.

Example 6.1

ST *Le visiteur **flâne** au bord du canal de l'Ourcq ou profite des nombreux espaces verts, bassins et fontaines.*
BT *The visitor **strolls** along the Canal de l'Ourcq or enjoys the many green spaces, ponds and fountains.*
Source https://www.parisinfo.com

Furthermore, the notion of reaching a destination is frequently conceptualised in Russian using the verbs *приезжать* (*arrive*) and *попадать* (*get to*), and in Greek with φτάνω (*arrive*) as illustrated in Examples 6.2 and 6.3, and less frequently in British English through the verb *reach*. Equivalents of these verbs have not been noted in the verbs frequently used in American English and French tourism texts. What is interesting about these verbs is that they are very general compared to some of the other Motion verbs. They encode no concrete information regarding how movement occurs, and they only focus on the result, which is displacement.

Example 6.2

ST *Попасть в Игарку можно только самолетом.*
BT *You can* ***get to*** *Igarka only by plane.*
Source https://www.russia.travel

Example 6.3

ST *Με σκάφος, με αυτοκίνητο, με τα πόδια θα* ***φτάσετε*** *στις πιο κοσμοπολίτικες, αλλά και στις πιο απομονωμένες [παραλίες].*
BT *By boat, by car, on foot you will* ***reach*** *the most cosmopolitan, but also the most isolated [beaches].*
Source www.discovergreece.com

Finally, British English and Greek tourism texts conceptualise movement in relation to a specific direction through the use of the verb *climb* and *ανεβαίνω* (*climb*) as illustrated in Examples 6.4 and 6.5. In this case, the focus is on the direction rather than on the effort associated with the movement and typically the goal is access to stunning views. Once again, equivalents of these verbs are not found in the other non-translated corpus components.

Example 6.4

After a short boat crossing from Sandwick, you can ***climb*** *the narrow steps to the top of the broch and take in breathtaking views across Mousa Sound.*
Source www.visitscotland.com

Example 6.5

ST *Καθώς* ***ανεβαίνετε*** *τα σκαλοπάτια, στο δρόμο που ξεκινάει από τα Χωράφια και φτάνοντας στην κορυφή του λόφου, σας περιμένει ένα πανόραμα σπάνιας ομορφιάς και ένα μοναστήρι-φρούριο: ο Αϊ-Γιώργης του βουνού!*
BT *As you* ***climb*** *the steps, on the road that starts from Chorafia and reaches the top of the hill, a panorama of rare beauty and a monastery-fortress awaits you: Saint George of the mountain!*
Source www.discovergreece.com

Even though movement is a process characterising all humans (with the possible exception of those with severe motion disorders), how much emphasis is placed on it as well as which aspects of motion are prioritised is culture-specific in these non-translated texts. Overall, Russian tourism discourse appears to place much more emphasis on Motion compared to English, while similarities are observed between French and Greek. At the same time, emphasis put on leisurely walks, reaching a destination, and direction of movement differs significantly, offering each language a unique profile. However, upward movement can be associated with terrain and landforms found at the destination—some might be flatter than others thus reducing the need for climbing—and therefore it is both space- and culture-specific. Using specific or generic descriptions of movement on the other hand is linked more to the language in which the text is written rather than what is being described. Russian and Greek tourism texts appear to conceptualise movement in more generic terms compared to English (both varieties) and French where more specific verbs, particularly those associated with a relaxed type of movement, can be found.

Based on the above, the Russian culture appears to be experiencing the sightseeing phase of tourism, while Anglophone cultures are in the post-sightseeing phase, placing less emphasis on typical features of tourist behaviour, such as motion, while also focusing on more nuanced elements of motion than simple displacement. French and Greek tourism texts are in-between these two extremes, with French potentially closer to the post-seeing phase than Greek.

6.3 Discovery

The second most popular semantic category overall is Discovery, which refers to finding a place or an object for the first time. In terms of the tourism phase, Discovery could be largely associated with the post-sightseeing phase, as it requires more than gazing at sights, although what aspects of it exactly are being highlighted plays an important role in gaining a clear picture of the tourism phase a society is experiencing.

As shown in Section 6.1, Greek tourism texts show the strongest preference for these verbs (31.54%). Next come French tourism texts (25.52%), closely followed by American (21.82%) and British English texts (21.55%). A very different picture is obtained from Russian tourism texts, where the frequency of Discovery verbs is strikingly low (7.29%). Statistical tests, conducted on the raw frequencies, reveal that differences are significant only when Russian tourism texts are compared to those in other languages, and when English (both varieties) tourism texts are compared to Greek texts. Therefore, a notable disfavour towards Discovery verbs is noted in Russian tourism texts, while a distinct preference is observed in Greek tourism texts when compared to English and, of course, Russian texts. This is an indication that Russian tourism texts are still experiencing the sightseeing phase, while tourism texts in the other languages, and most notably Greek texts, appear to have moved away from it. However, a closer inspection of which verbs realise this category is needed.

Discovery appears to consist of three main elements: looking for something, exploring the destination for it, and finding it. Verbs encoding these three elements are frequently used in all languages examined here apart from French and Russian. In French the element of looking for something is missing, while in Russian only that of finding is present.

Looking for something is realised through the verbs *look for* in British and American English and *αναζητώ* (*look for*) in Greek as illustrated in Examples 6.6 and 6.7. The focus is on tourist needs often prior to travelling, which the tourism text tries to capitalise on to increase its 'pull' factor and trigger the desired response. The destination, or specific elements of it, is presented as the (best) answer to the tourist's quest. While some plausible explanations linked to the tourism phase can be put forward regarding why this element is missing from Russian tourism discourse, such as that societies in the sightseeing phase focus less on discovery, it is less clear why French tourism texts do not frequently use verbs such as *chercher* (*look for*). However, as this element is not very prominent in any of the other languages in which it has been observed, this might simply be due to a low occurrence which means that instances have not been observed in the data as a result of focusing on the most frequent verbs.

Example 6.6

Looking for *a quite [sic] spot among the West End's hustle and bustle?*
Source www.visitlondon.com

Example 6.7

ST *Αν αυτό που* ***αναζητάτε*** *στις διακοπές σας είναι ώρες ξεγνασιάς, τότε η Κάρυστος είναι ο ιδανικός προορισμός για εσάς.*
BT *If what you are* ***looking for*** *on your vacation are hours of relaxation, then Karystos is the ideal destination for you.*
Source www.discovergreece.com

The second element, that is exploration, plays a much more central role in the conceptualisation of this semantic category. Exploration is realised through the verbs *explore* in British and American English, *explorer* (*explore*) in French, and *εξερευνώ* (*explore*) in Greek. What can be explored is normally limited to nature (Example 6.8), urban settings (Example 6.9), and occasionally to establishments connected to history and civilisation (Example 6.10).

Example 6.8

Explore *a scenic wildlife refuge teeming with migratory birds, or take a fishing trip or a paddleboard exploration through Delaware's peaceful inland bays.*
Source www.visittheusa.co.uk

Example 6.9

ST ***Explorez*** *les hauteurs de la ville pour découvrir la citadelle de Sisteron, veritable couronne de pierre.*
BT ***Explore*** *the heights of the city to discover the citadel of Sisteron, a veritable crown of stone.*
Source www.provence-alpes-cotedazur.com

Example 6.10

ST *Μια ευκαιρία να* ***εξερευνήσετε*** *το μοναδικό Μουσείο που διηγείται τη γέννηση και την εξέλιξη του Φιλελληνισμού και την επίδραση που έχει ο Ελληνικός Πολιτισμός στον κόσμο, μέσα από έργα και αντικείμενα τέχνης, έγγραφα και τεκμήρια, από την Αναγέννηση μέχρι σήμερα.*

BT *An opportunity to* ***explore*** *the unique Museum that tells the story of the birth and development of Philhellenism and the impact that Greek Culture has on the world, through works and art objects, documents and evidence, from the Renaissance to the present day.*

Source www.visitgreece.gr

What makes exploration an interesting element of Discovery verbs is that it involves a certain element of uncertainty while it also demands effort on behalf of the tourist. In the examples above, the outcome of each exploration might be different for each tourist, as there are seemingly endless options even within the limitations of a specific space. Additionally, tourists need to actively engage with the exploration and what the destination has to offer. Both these elements are in striking contrast to the mainstream idea of tourism associated with sightseeing. A society that is still experiencing the sightseeing phase of tourism is likely to avoid any references to exploration, which is what has been observed in Russian tourism texts. If any Discovery verbs are used, they are more likely to belong to the third and final element, that is the outcome of the exploration.

Finding what one is looking for at a destination is mainly realised by *discover* and *find* in British and American English, *découvrir* (*discover*) in French, *ανακαλύπτω* (*discover*) and *βρίσκω* (*find*) in Greek and *находить* (*find*) in Russian. A very wide range of things can be discovered or found, as can be seen in Examples 6.11 and 6.12, suggesting that discovery is interwoven with all aspects of tourism. Focusing on finding instead of exploring makes the tourism experience much more predictable in the eyes of the reader-tourist. The outcome of the exploration is predetermined, and although options exist, as evidenced in the examples, these constitute a finite set.

Example 6.11

The small towns like Chester, Guilford, Essex and Kent offer big experiences, where you'll. ***Discover*** *shopping, dining and culture to fill a day or a weekend.*
Source www.visittheusa.co.uk

Example 6.12

ST *Στον Πλατανιά θα* ***βρείτε*** *έντονη νυχτερινή ζωή, μπαρ, εστιατόρια, μια μεγάλη αμμώδη παραλία και αμέτρητα ξενοδοχεία που απλώνονται στην παραλία του.*
BT *In Platanias you will* ***find*** *lively nightlife, bars, restaurants, a long sandy beach and countless hotels that line its beach.*
Source www.incrediblecrete.gr

Since more verbs associated with the exploration and finding elements have generally been found, it is interesting to compare these two elements more closely. In British and American English tourism texts both are equally distributed with approximately the same number of verbs associated with each element. English tourism discourse encodes a notion of tourism where predetermined discoveries, as well as more open-ended opportunistic explorations are favoured. Both French and Greek tourism texts show a much stronger preference towards the outcome of the exploration, referring to the exploration itself only occasionally. There is perhaps some evidence that French and Greek tourism discourses try to move away from the sightseeing phase of tourism. Finally, Russian tourism discourse encodes a notion of tourist behaviour which is rather mainstream and where exploration is silenced. What we have here is British and American culture clearly experiencing the post-sightseeing phase, while Russian culture is still in the sightseeing phase. French and Greek stand somewhere in between with evidence that the next phase might have already started, more so in Greek perhaps considering also the high frequency of Discovery verbs compared to other languages, even English.

While Discovery is a key feature of tourism discourse in all languages examined, the focus on the different elements of this semantic category is culture-specific. Overall, Russian tourism discourse appears to place less emphasis on Discovery, focusing only on what can be found at a destination. On the other hand, Greek tourism texts appear to place a strong emphasis on this category, but this is mitigated by focusing more strongly on the outcome of the exploration rather than the process of exploration itself, which has also been observed with French tourism texts. At the same time, English tourism texts show a clear picture of favouring Discovery and focusing on all elements of it, including most importantly open-ended exploration.

Although the picture obtained from the semantic category of Discovery is not as clear as that obtained from Motion verbs, there is once again evidence to suggest that Russian society is experiencing the sightseeing phase of tourism, while Anglophone societies are in the post-sightseeing phase, paying more attention to individual explorations. French and Greek tourism texts are in-between these two extremes, with Greek potentially closer to the post-sightseeing phase than French.

6.4 Perception

While the previous two semantic categories capture tourist behaviour that is related to active participation, Perception is related to sensations, and therefore requires little active involvement. The prominence of this semantic category is not surprising, especially since vision is strongly associated with tourism. By understanding how much a society focuses on Perception, and which senses are being highlighted, it is possible to gain a better picture of whether it still experiences the sightseeing phase, which is naturally strongly associated with vision.

The strongest preference towards Perception is found in Russian tourism texts (33.37%), by far exceeding the frequencies of all other languages. The second highest frequency is found in Greek tourism texts (23.09%), closely followed by British English (21.96%), French (19.47%), and finally American English tourism texts (19.09%). Statistical tests, conducted on the raw frequencies, reveal that differences are significant only when Russian tourism texts are compared to texts in any other language. Therefore, there is a very notable preference for Perception verbs in Russian tourism discourse. This is an indication that the

Russian culture is experiencing the sightseeing phase of tourism much more than any other society examined here. The prominence of Perception, and in particular sight, in Russian tourism texts is exemplified in Example 6.13, where the phrase *своими глазами* (*with their own eyes*) is added for emphasis.

Example 6.13

ST *Туристы могут пройтись возле строений того времени,* ***посмотреть своими глазами*** *на дома, украшения и оружие эллинов.*
BT *Tourists can walk near the buildings of that time,* ***see with their own eyes*** *the houses, decorations and weapons of the Hellenes.*
Source www.russia.travel

Vision-related verbs are only one of the ways through which Perception is expressed. Other types of verbs include verbs related to taste, and experience-related verbs. These are generally found in all languages with some notable exceptions, however, which are discussed below.

Vision-related verbs include *see, watch, spot* in English, *voir* (*see*), *admirer* (*admire*), *observer* (*observe*) in French, *βλέπω* (*see*) and *θαυμάζω* (*admire*) in Greek, and *увидеть* (*see*) and *посмотреть* (*look*) in Russian. The fact that all languages emphasise vision is not surprising, considering how strongly tourism is (still) associated with gaze (see Case study II and specifically Chapter 8 for a more detailed discussion about tourist gaze). What can be perceived through the eyes involves a wide range of elements, such as sights (Example 6.14), artefacts (Example 6.15), or landscape (Example 6.16).

Example 6.14

ST *Μην ξεχάσετε να **δείτε** τα αξιοθέατα: τον Καθεδρικό ναό του Τιμίου Προδρόμου του 1838, το Τελωνείο, την Ψαραγορά, το πέτρινο γεφύρι, το άγαλμα του μικρού ψαρά, το ρολόι του 1881 και τον παλιό ταρσανά στη συνοικία Χαράνι.*

BT *Do not forget to **see** the sights: the Cathedral of the Holy Forerunner of 1838, the Customs House, the Fish Market, the stone bridge, the statue of the little fisherman, the clock of 1881 and the old tarsana in the Harani district.*

Source www.discovergreece.com

Example 6.15

ST *Среди экспонатов Зала Славы можно **увидеть** портреты, фотографии, награды, кубки, медали, этикетки, музейные артефакты.*

BT *Among the exhibits of the Hall of Fame you can **see** portraits, photographs, awards, cups, medals, labels, museum artifacts.*

Source www.russia.travel

Example 6.16

*Around the monument is the glorious surroundings of Loch Shiel which makes for a perfect walk to **admire** the landscape in all its glory.*

Source www.visitscotland.com

It is interesting that *admire*, which can be defined in the context of tourism as looking at something with pleasure, is not found in American English and Russian tourism texts—an interesting similarity between two otherwise very distinct tourism discourses. What seems to be missing is the element of pleasure—this will be further examined when the Emotion category is discussed later.

While vision appears to be a shared element of tourism discourse across languages, no such equal emphasis is placed on verbs related to taste. Taste-related verbs include *try* and *sample* in English, *déguster* (*taste*) and *goûter* (*taste*) in French, *δοκιμάζω* (*try*) in Greek, and *попробовать* (*try*) in Russian. These verbs are more frequently used in French tourism texts accounting for approximately 25% of all Perception verbs, while French is also the only language where not just one, but two verbs with a meaning limited to tasting with the mouth are used. This reflects the importance that French people assign to good food and the popularity of French cuisine.

On the other hand, taste-related verbs are not found in the American English tourism texts, while their frequency out of all Perception verbs is very low (7%) in British English texts. American and British cuisines are not famous for high-quality food, which means that destination promoters are unlikely to promote them. However, what they are more likely to promote in the case of British tourism is alcoholic drinks (Example 6.17) and tours around distilleries and breweries are frequently mentioned. Beer, gin, and whisky are presented as high-quality drinks which form part of the British tourism experience.

Example 6.17

*Make sure you **sample** some of Plymouth's famous gin at the Black Friars Distillery.*
Source www.visitbritain.com

While alcoholic drinks, mostly wine, are also an integral part of French and Greek tourism discourse and are often mentioned in tourism texts, more emphasis is placed on traditional and representative dishes, as well as dishes that the visitor is unlikely to have ever tried at home (Examples 6.18 and 6.19).

Example 6.18

ST *__Δοκιμάστε__ το ντόπιο θυμαρίσιο μέλι αρίστης ποιότητας, το μαλακό τυρί που παρασκευάζεται από αιγοπρόβειο γάλα με παραδοσιακό τρόπο, ολόγλυκα φραγκόσυκα, τα ντόπια χειροποίητα μακαρούνια (μακαρόνια) και κουλιά (ζυμαρικό σε σχήμα κοχυλιού), αρνάκι ή κατσικάκι οφτό (γεμιστό) μαγειρεμένο σε «ξυλόφουρνο» (το πιο "γιορτινό" φαγητό), τα μικρούτσικα πεντανόστιμα γιαπράκια (ντολμαδάκια με κιμά), φρεσκότατα ψάρια (γευθείτε τη νοστιμότατη κακαβιά, σκάρο τηγανητό, και Χαλκίτικο γαριδάκι που τρώγεται ολόκληρο μόλις τηγανιστεί).*

BT ***Try** the local high-quality thyme honey, the soft cheese made from goat's milk in the traditional way, all-sweet prickly pears, the local handmade makarounia (pasta) and koulia (shell-shaped pasta), lamb or goat ofto (stuffed) cooked in a wood-fired oven (the most "celebratory" food), the small, delicious giaprakia (vine leaves stuffed with minced meat), fresh fish (taste the delicious kakavia, fried parrotfish, and Chalki's shrimp that is eaten whole once fried).*

Source www.visitgreece.gr

Example 6.19

ST *À Marseille, on **déguste** la bouillabaisse, soupe de poissons emblématique et à Martigues, la poutargue, " caviar " local, d'œufs de mulet ou de thon rouge séchés.*

BT *In Marseille, you can **taste** the bouillabaisse, emblematic fish soup and in Martigues,the poutargue, local "caviar", made from mullet roe or dried bluefin tuna.*

Source www.france.fr

While acknowledging that cuisine—regardless of its particular reputation—is something very strongly associated with each culture and location and therefore it is expected that differences will be noted both in terms of what and how much it is emphasised, it is important to consider that taste is a key element of the tourism experience, as also demonstrated by the recent *Visit Britain* survey (Hickey et al. 2022) where food and drink options are important drivers for Western European and North-East Asian tourists. This suggests that it is not simply that French tourism texts have more to say about local food and drink, but also, and perhaps more importantly, that French-speaking visitors are assumed to associate tourism with gastronomy. This finding is not so much related to the tourism phase that a culture might experience but rather to individual cultural preferences.

Finally, some Perception verbs are not associated with a particular sense but refer to sensations more generally. These include *experience* and *feel* in English, *vivre* (*live*) in French, and *ζω* (*live*) in Greek. The reason why such verbs are classified as Perception verbs is because they are closely linked to sensations, as in Example 6.20 where *experience the sights and sounds* refers to vision and sound. Russian tourism texts do not appear to employ such verbs frequently and instead prefer to focus on a specific sense.

Example 6.20

*A great way to **experience** the sights and sounds of Little Havana at night is during Viernes Culturales/Cultural Fridays – the neighborhood's monthly arts and cultural event.*
Source www.miamiandbeaches.com

The decision to include *vivre* (*live*) in French, and *ζω* (*live*) in Greek under Perception verbs is based on the fact that although strictly speaking they are back-translated as *live*, in the tourism context their meaning is much closer to that of *experience*, as illustrated in Example 6.21, and like *experience*, they involve all the senses.

Example 6.21

ST *Τα νησιά του Ιονίου και του Αιγαίου και οι ακτογραμμές της ηπειρωτικής Ελλάδας σας περιμένουν, για να **ζήσετε** το ελληνικό καλοκαίρι και να χαρείτε τον ήλιο καιτη θάλασσσαμαζί με τους αγαπημένους σας.*
BT *The islands of the Ionian and Aegean and the coastlines of mainland Greece are waiting for you, to **experience** the Greek summer and enjoy the sun and the sea with your loved ones.*
Source www.discovergreece.com

While it might be easy to dismiss these experience-related verbs as simply expressing Perception in more general terms, they are in fact extremely important when trying to understand how tourist behaviour is conceptualised. According to O'Dell (2005, 12) "[e]xperiences have become the hottest commodities the marker has to offer" and he coins the term *experiencescapes* to refer to landscapes of experience. In the Western world, he argues, more and more aspects of leisure and tourism are packaged and communicated as experiences. Tourists no longer want to simply passively observe things, they want to engage with them and immerse themselves more fully in the local culture. This is further supported by the report commissioned by Oxford Economics and Amadeus (2018, 29) which identifies a trend from destination-based to experience-based travel. This trend is typically associated with the preference for more active holidays and not necessarily with a subtle, yet meaningful shift, regarding how sensory experiences might be conceptualised. Further evidence of this trend can be found in how the aims of Marketing Greece, which is responsible for *Discover Greece* are defined on the website: "It aims to improve the characteristics of the demand for holidays in Greece. In other words, upgrading in the qualitative aspects of Greek tourism, chiefly by promoting the *experiential aspect of holidays in Greece*" (*Discover Greece*, About Us page, emphasis added). It is precisely in the use of the verb *experience* and its possible equivalents that we can see the shift from mainstream into alternative tourism, from passive sightseeing to active engagement—it is the space where actions and sensations merge into one.

While the importance of experiencing a destination cannot be overstated, what previous research suggests is that this shift towards experience-based tourism is somehow universal. Some evidence that it might not be is provided by Sulaiman and Wilson (2019) but their study involved the comparison of Australia and Malaysia—two very distinct societies. Data from the TrAIL corpus point towards cross-cultural differences even within what might be perceived as the Western world. Experience-related verbs are more frequently found in American English tourism texts (21% of all Perception verbs), followed by British English (14%), French (10%), and Greek (9%). There is a clear indication here that this shift is much more pronounced in Anglophone societies, and even more so in the American one, while it is not observed in the Russian society, which is part of the eastern world. As with many of the other categories examined so far, French and Greek tourism texts stand in-between these two extremes.

The senses naturally play a central role in the tourism experience, but how much emphasis is placed on them, and which senses exactly are prioritised is found to be culture-specific. Overall, Russian tourism discourse appears to place much more emphasis on the senses, and particularly vision, compared to the other languages examined here. At the same time, English tourism texts show a much stronger preference for experience-related verbs, which are absent in Russian tourism discourse and only occasionally found in French and Greek texts.

Based on the above, there appears to be further evidence to support the idea that Russian society appears to be experiencing the sightseeing phase of tourism, while Anglophone cultures are in the post-sightseeing phase, promoting experience-based travel. French and Greek tourism texts are in-between these two extremes.

6.5 Emotion

The final category to be examined is that of Emotion, which broadly refers to human feelings, which like Perception, requires little active involvement. Unlike the semantic categories examined so far, it is not easy to associate Emotion with a tourism phase and the discussion that follows reveals that this category is associated with a middle point on the continuum. Additionally, being the least prominent category means that it is also important to consider results in conjunction with what has already been observed for the other semantic categories.

The strongest preference towards Emotion verbs is found in French (12.14%) and Greek (12.11%) tourism texts. Next come American tourism texts (9.89%), closely followed by British English tourism texts (8.93%), while Emotion verbs have not been found as being frequently found in Russian tourism texts. Statistical tests, conducted on the raw frequencies, reveal that differences are statistically significant only when British English texts are compared to French and Greek texts, or in other words, no statistically significant differences were found between French and Greek tourism texts or tourism texts in different varieties of English. Therefore, a preference for Emotion verbs is clearly noted in French and Greek tourism texts, while a notable disfavour towards them is observed in Russian ones. English texts stand in-between but closer to French and Greek than Russian.

These findings further confirm that there is evidence from Russian tourism discourse that the Russian society is still experiencing the sightseeing phase of tourism and has not yet reached the stage where attention is also paid to enjoyment. At the same time, evidence from British and American tourism texts suggest that Anglophone societies have reached the post-sightseeing phase and while they promote aspects of tourism other than enjoyment, some attention is also paid to this emotion. This relative low emphasis on enjoyment in the Russian tourism texts might also explain the absence of an equivalent of *admire* and *stroll* in the Russian tourism texts. Finally, French and Greek cultures, as also noted earlier, appear somewhere between these two extremes—somewhere in the middle of their journey towards the post-sightseeing phase. It appears that enjoyment is mainly associated with the middle point in the continuum of tourism phases and this is why it presents such a unique profile.

Regarding verb choices, it is notable that *enjoy* and its equivalents are used in all languages where the semantic category of Emotion has been noted, which makes its absence from Russian tourism texts even more noteworthy. Another similarity across languages is that by far the most frequent aspect of the destination that can be enjoyed is the *view* (Example 6.22).

Example 6.22

ST *Du parc, vous* ***profiterez*** *d'une vue imprenable sur la mer et pourrez vous détendre àl'ombre des arbres ou entamer une partie de pétanque.*

BT *From the park, you will* ***enjoy*** *a breathtaking view of the sea and can relax in the shade of the trees or start a game of pétanque.*

Source www.provence-alpes-cotedazur.com

Other aspects that can be enjoyed include food, strolls, music, beaches, holidays, and nature. Interestingly, a common collocation of *enjoy* only found in American English is *beauty* as illustrated in Example 6.23. In this case, pleasure can be derived from abstract things which are very hard to define, unlike food or music, for instance. Focusing on beauty can be perceived as an expression of the desire to experience inner peace and harmony, providing further evidence that American society has reached the post-sightseeing phase, focusing more on self-fulfilment.

Example 6.23

> *While the lake is certainly the main attraction, you'll also find plenty of opportunities to* ***enjoy*** *the quiet, stark beauty of the surrounding Mojave Desert.*
> *Source* www.visittheusa.com

A similar observation can be made about *soak up* which is only found in British English tourism texts. The most frequent collocations of this verb are *atmosphere* and *views*. Like beauty, atmosphere is rather abstract and hard to define. For instance, a *literary atmosphere* (Example 6.24) is an illusive concept. Only Anglophone societies appear to associate pleasure with intangible aspects of the destination.

Example 6.24

> *At the Robert Burns Birthplace Museum you can* ***soak up*** *the literary atmosphere among the collection of his works before exploring Burns Cottage where the poet was born.*
> *Source* www.visitbritain.com

Although Emotion is not as strongly associated with tourism as the other semantic categories examined so far, its examination provides useful insights into how tourism behaviour is conceptualised across languages. Overall, a clear disfavour towards tourist emotions is noted in Russian tourism texts, while the French and Greek texts appear to favour this category the most. At the same time, while English tourism texts do not appear to rely heavily on such verbs, whenever they do, they also associate them with more abstract concepts, something that has not been observed in any of the other languages examined here.

By examining patterns in conjunction with each other, it is evident that while Anglophone societies appear to have moved into the post-sightseeing phase, Russian tourism culture is still firmly grounded in the sightseeing phase. French and Greek tourism texts are in-between these two extremes.

6.6 Conclusions

The examination of the most prominent semantic categories of verbs where the tourist is the explicit or implied subject/agent has provided profound insight into how each language examined here perceives and conceptualises tourist behaviour, and the notion of tourism more generally. This has provided valuable insights into the tourism phase each society is currently experiencing. Using the notion of the tourism phase as an analytical framework, it has been possible to assign each society to different points in the continuum ranging from the sightseeing phase to the post-sightseeing phase. The linguistic elements associated with the two extremes of the continuum are summarised in Table 6.3 (items in brackets indicate features that are not particularly strongly associated with either phase). It is important to stress once again that this is not a binary opposition. In other words, it does not necessarily mean that an element that is representative of the sightseeing phase is totally absent from the post-sightseeing phase—it is rather a matter of degree.

Societies experiencing the sightseeing phase have a very mainstream view of tourism and focus on its most prototypical elements, such as movement through space and viewing attractions, while those in the post-sightseeing phase have a more adventurous view of tourism and are likely to promote experiencing a destination and active exploration. Based on the data obtained from the analysis of the non-translated corpus components, the five societies examined here can be mapped onto the continuum as illustrated in Figure 6.2 below. It is worth noting that

Table 6.3 Linguistic elements associated with each phase of tourism

		Sightseeing phase	Post-sightseeing phase
Motion	*Movement*	X	
	Specific movement		X
Discovery	*Exploration*		X
	Outcome	X	
Perception	*Vision*	X	
	Experience		X
Emotion	*Pleasure*	(X)	(X)
	Abstract elements		X

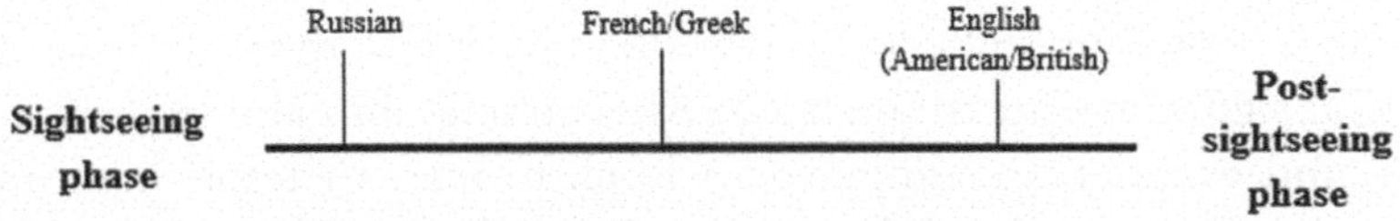

Fig. 6.2 TrAIL corpus non-translated components tourism phase continuum

results from the examination of different semantic categories *consistently* point to the mapping below, therefore increasing the confidence of the analytical model and the reliability of the findings.

At the one end of the continuum we find Russian tourism discourse, which clearly reflects a mainstream view of tourism grounded in the sightseeing phase. Russian tourism texts present a view of tourism which mainly consists of travelling to a place to look at sights. What can be found at a destination needs to be predetermined and there is generally little risk or active involvement associated with travelling. Pleasure is not something that is considered an important element of the tourism experience according to Russian tourism discourse.

At the other end of the continuum are American and British English tourism discourses, which have embraced the post-seeing phase of tourism as reflected in the more experiential view of tourism that is encoded in these tourism texts. English tourism texts (at least American and British) conceptualise tourism as an active endeavour in which reader-tourists play a central role. Emphasis is put on open-ended exploration where tourists can make their independent discoveries, as well as on engaging with the destination on a more meaningful level than simply gazing at sights. Travelling to a destination means exploring and experiencing it, with only some attention being paid to pleasure. It is interesting that no significant differences have been observed between the two varieties of English across the semantic categories, which might justify the decision of tourism websites not to localise in terms of a specific variety of English, but rather rely on an international version of it.

At the middle point of the continuum, we find French and Russian tourism discourses, which have started moving away from the mainstream view of tourism and embracing more the post-sightseeing view, although they have still not reached the level occupied by English. French and Greek tourism texts focus less on movement through space and gazing at sights, at least compared to Russian ones, but also less on exploration and experiences compared to English ones, therefore combining elements of both ends of the continuum. What sets apart French and Greek tourism texts from the rest is the attention paid to pleasure, which is associated with the mid-point in the continuum. However, some subtle differences are noted even between French and Greek tourism texts; for instance, a much stronger emphasis is put on Discovery verbs in Greek texts, or more specific verbs are used to conceptualise movement in French texts. These differences point at times towards the sightseeing and at times towards the post-seeing phase, further indicating that these two cultures are in the process of moving towards the next tourism phase.

Mapping languages in this way reveals that the much used (and overused) label of 'western world', often found in tourism studies (see, for example, Amadeous and Oxford Economics 2018) can be misleading at least when it comes to tourism. While Russia does not belong to the Western world (at least not as clearly as the other countries examined here do) and therefore differences between Russian and English tourism discourse might have been anticipated, both France and Greece belong to the Western world and also have very strong tourism industries. Yet, the conceptualisation of tourist behaviour according to French and Greek tourism texts is different to that found in English, revealing that even within the so-called Western world the perceptions of tourism change at different speeds. Something similar might be expected of the eastern world and there is no reason to assume that the Russian perception of tourism will be shared by the Chinese or Japanese societies or indeed societies of the Middle East, even though similarities are bound to exist—for instance, Sulaiman and Wilson (2019) argue that Malay culture is also still experiencing the sightseeing phase. This highlights the importance of avoiding overgeneralisations and obtaining the individual profile of each society and language based on a linguistic analysis of its communicative preferences.

Insight into how each language and culture conceptualises tourist behaviour, and tourism more generally, can inform the translation process, as it can dictate the amount and type of adaptation required when tourism texts are translated into other languages. When translation occurs into English, which is the most typical scenario, adaptation will be required to bring out elements associated with the post-sightseeing phase, while the reverse will be required when translation occurs out of English. Similarly, high degrees of adaptation are expected when translating tourism texts into/out of Russian. Lack of adaptation, when required, can potentially have a profound effect on the success of the tourism message. A text that addresses an audience that is still experiencing the sightseeing phase highlighting elements associated with the post-sightseeing one, such as active exploration and experience-based tourism, is likely to intimidate readers and make them feel reluctant about visiting that destination. Conversely, a text which reflects the sightseeing phase and addresses an audience that has moved further away from it, might appear uninteresting and unexciting. Whether or not translators manage to strike the right balance and successfully negotiate the differences in the way tourism behaviour and tourism is conceptualised across languages and cultures will be examined in the next chapter.

References

Dann, Graham M.S. 1996. *The Language of Tourism.* Oxon: CAB International.

Francesconi, Sabrina. 2007. *English for Tourism Promotion: Italy in British Tourism Texts*. Milano: Hoepli.

Hickey, K., N. Chan, and C. Warren. 2022. "MIDAS: A Global Report—Motivations, Influences, Decisions and Sustainability in a Post-Pandemic Era."

O'Dell, T. 2005. "Experiencescapes: Blurring Borders and Testing Connections." In *Experiencescapes: Tourism, Culture and Economy*, edited by T. O'Dell and P. Billing, 11–33. Koge: Copenhagen Business School Press.

Netto, Panosso, and Alexander. 2009. "What Is Tourism? Definitions, Theoretical Phases and Principles." In *Philosophical Issues in Tourism*, edited by John Tribe, 43–61. Bristol: Channel View Publications.

Amadeus and Oxford Economics. 2018. "Shaping the Future of Travel: Macro Trends Driving Industry Growth over the next Decade."

Sulaiman, M. Zain., and Rita Wilson. 2019. *Translation and Tourism*. Singapore: Springer.

7

Translating Tourist Behaviour

In the previous chapter we saw how the languages examined as part of the TrAIL corpus differ regarding their conceptualisation of tourist behaviour, and tourism more generally, as well as the tourism phase they are currently experiencing. Now that the differences have been established, the next step is to examine how they are negotiated in translation, and to what extent translations of tourism texts successfully (re)present what the destination has to offer while accommodating for the worldview of the reader. To facilitate the analysis, the discussion is organised around the translated languages: English, French, and Russian. In each section, comparisons are made between target texts and both their source texts and non-translated texts in the target language.

Adaptation is examined mainly in relation to the frequency distribution of the semantic categories, and also in terms of how specific elements of the categories identified in the previous chapter as important are treated during translation. The focus is on the four most prominent categories identified in the previous chapter, that is, Motion, Discovery, Perception, and Emotion. Translations that attempt to come close to non-translated texts in the target language, while at the same time recognising where adaptation is not required, are considered to successfully

S. Malamatidou, *Translating Tourism*, Palgrave Studies in Translating and Interpreting, https://doi.org/10.1007/978-3-031-49349-2_7

bridge the gap between source and target society and to (re)create an image of tourism that is in line with how reader-tourists are likely to experience the world, while at the same time presenting what the destination has to offer. The success of the adaptation is also likely to affect the promotional potential of these texts, creating a more positive image of the destination, and turning more readers into visitors.

It is worth recalling that adaptation is understood here as changes that lead to a better fit and go beyond the meaning of words (see Chapter 3 for a more detailed discussion). The degree and type of adaptation is established based on statistical significance tests, as explained in Table 4.4. For example, whenever successful adaptation is mentioned that means that the following two criteria have been met. To recap, the first is a comparison of the non-translated texts in the relevant two languages that shows them to be different at a statistically significant level, indicating the need for adaptation in translation. The second comparison is between the non-translated texts now serving as source texts and their target texts; if adaptation has been successful, a statistically significant difference will also be noted here, while no statistically significant difference will be noted between the target texts and the non-translated texts in that language. For ease of reference, and to avoid a number-heavy reporting of findings, in this section and the rest in this chapter, only normalised frequencies of the verbs assigned to the various semantic categories will be reported, and results of statistical significance tests, provided by the online calculator, are implied in the adaptation categories reported.

7.1 Tourism Texts Translated into English

For the examination of translated English, no distinction is being made between British and American English for translated texts, since the variety of English is typically not specified on the tourism websites examined as part of the TrAIL corpus. A version of international English seems to be employed, whatever this might mean linguistically. Additionally, no significant differences have been observed between American and British English texts during the previous stage of analysis, reflecting that both societies are experiencing the same phase of tourism, i.e. post-sightseeing, which further supports the idea that it is possible to talk about a single English version which can address both the American and the British audience.

To establish the basis for the calculation of the normalised frequency of verbs belonging to specific semantic categories, the first step is the examination of the overall distribution of verbs which meet the threshold frequency measure of 0.18ptw and which also have the reader-tourist as their subject/agent in non-translated texts and in translated texts and their source texts, which is summarised in Table 7.1. Although not the focus of the present exploration, it is interesting to note that target texts generally stay close to their source texts regarding the frequency distribution of verbs.

The next step is to examine the distribution of the four most prominent semantic categories to reveal adaptation patterns. Normalised frequencies are presented in Table 7.2.

Table 7.3 summarises adaptation patterns. Overall, as can be seen, languages show very varied patterns when it comes to individual semantic categories. Even within the same language pair, different semantic categories are treated differently when it comes to adaptation, but in most cases, it is possible to identify a general trend based on the type of adaptation most categories belong to. As explained in Chapter 4, these conclusions are based on log-likelihood and Bayes Factor (BIC)

Table 7.1 Normalised frequencies of most frequent verbs with reader-tourist as subject/agent in non-translated English, source texts (i.e. non-translated texts in French, Greek, Russian), and translated English corpus components

		Tokens	Verbs (types)	Verbs (tokens)	
				f	nf (ptw)
Non-translated texts	EN_US	170,520	35	2,750	16.12
	EN_UK	170,704	53	5,082	29.77
Source texts	FR	170,060	27	2,018	11.86
	EL	170,601	20	1,503	8.81
	RU	170,881	14	809	4.73
Translated texts	FR-EN	149,421	31	2,296	15.36
	EL-EN	170,369	20	1,660	9.74
	RU-EN	186,169	9	768	4.12

Table 7.2 Normalised frequencies of most prominent semantic categories of verbs in non-translated English, source texts (i.e. non-translated texts in French, Greek, Russian), and translated English corpus components as a percentage of identified verbs (tokens)

		NT TEXTS (TL)		NT TEXTS (SL)			TRANSLATED TEXTS		
		EN_US	EN_UK	FR	EL	RU	FR- EN	EL- EN	RU- EN
Verbs (tokens)		2,750	5,082	2,018	1,503	809	2,296	1,660	768
Motion	Raw frequency	646	1,011	535	394	320	630	597	323
	Normalized frequency	23.49%	19.89%	26.51%	26.21%	39.56%	27.44%	35.96%	42.06%
Discovery	Raw frequency	600	1,095	515	474	59	482	378	104
	Normalized frequency	21.82%	21.55%	25.52%	31.54%	7.29%	20.99%	22.77%	13.54%
Perception	Raw frequency	525	1,116	393	347	270	392	211	198
	Normalized frequency	19.09%	21.96%	19.47%	23.09%	33.37%	17.07%	12.71%	25.78%
Emotion	Raw frequency	272	454	245	182	0	305	211	53
	Normalized frequency	9.89%	8.93%	12.14%	12.11%	0%	13.28%	12.71%	6.90%

statistical comparisons of the raw frequencies of verb occurrences in pairs of non-translated texts and the translated texts in the relevant language, and of translated texts and their source texts (to compare the use of verbs in each TT with that in the source locale). A detailed presentation of what each type of adaptation refers to can be found in Chapter 4 (Table 4.4)

Even though there is generally a need for adaptation due to the differences between non-translated English tourism texts and those in the other languages examined here, as evidenced in Chapter 6, very few instances of successful, or even weak, adaption are observed in the translations. Additionally, there is a lot of variation in terms of how translators have approached the tourism texts, ranging from adaptation even when it is not required to no adaptation. Overall, each language pair appears to have a different preference, with French-English staying close to the source texts, and Russian-English demonstrating the most concerted attempt towards adaptation, while Greek-English does not offer a clear pattern of adaptation. Translations into English are the only set that presents such a varied picture depending on the language pair. To examine semantic categories in more depth and what effect the different degrees and types of adaptation have on the discourse of tourism, the discussion will now focus on individual language pairs.

Table 7.3 Patterns of adaptation observed in translations into English (verbs)

	FR-EN	EL-EN	RU-EN
Not Required	Motion Discovery Perception	Emotion	
Additional		Motion Perception	
No	Emotion		Motion
Successful		Discovery	Emotion
Weak			Discovery Perception
Over			
Reverse			

7.1.1 Translations into English from French

When French tourism texts are translated into English, a generally low degree of adaptation might be expected as a result of the similarities between French and English non-translated texts regarding the distribution of Motion, Discovery, and Perception verbs (these have been established in Chapter 6 based on the statistical tests for each semantic category). Indeed, adaptation is not observed for these semantic categories. However, adaptation might be expected in the case of Emotion verbs, since a statistically significant difference has been observed in their frequency in English and French non-translated tourism texts, but is not observed in the translations. Overall, no instances of successful or weak adaptation are noted in translations of French tourism texts into English. Figure 7.1 summarises the quantitative findings presented earlier in Table 7.2.

It is worth recalling that in this, and all similar figures in the rest of this chapter, only the four most prominent semantic categories of verbs, identified in Chapter 6, are presented and discussed. However, their frequency is measured out of the total number of verbs identified (see Table 7.1), which explains why they do not add up to 100%.

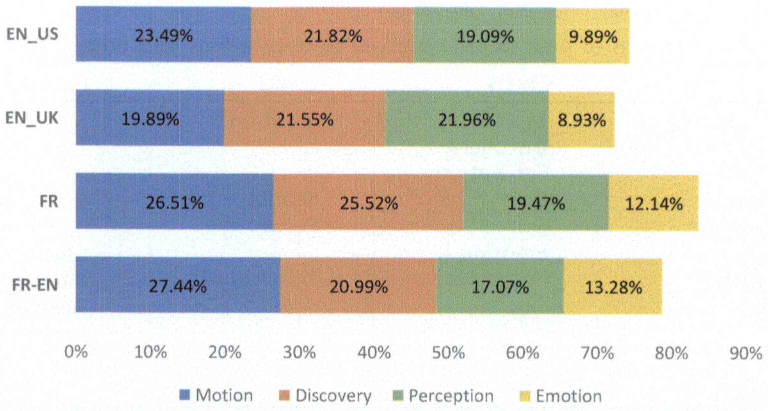

Fig. 7.1 Normalised frequencies of most prominent semantic in non-translated English, French source texts and translated English from French as % of all identified verbs

It might be easy to assume that since no adaptation is required in the case of certain semantic categories based on the results from the statistical tests conducted in Chapter 6, and none is observed, that means that the resulting translated text is successful in promoting the notion of tourism behaviour that is in accordance with the target readership's worldview. However, the previous stage of finer-grid analysis (see Sects. 6.3 and 6.4) revealed some important differences between English and French regarding the exploration/outcome focus of Discovery verbs, and the experience element of Perception verbs. Examination of the French translations into English suggests that no attempt has been made to adapt these to match what is found in English non-translated texts with regard to these elements. Particularly noteworthy is the fact that although *vivre* (*live*) is found in French tourism texts, its English equivalent *experience* is not found in their translations. Considering how much emphasis is put on experiences in the post-sightseeing phase of tourism, this is a notable omission with potentially significant consequences.

No adaptation is observed for the semantic category of Emotion. Although Emotion verbs represent a lower frequency of all verbs in both American (9.89%) and British English (8.93%) tourism texts than French texts (13.28%), target texts stay close to their source texts (12.14%). This means that more emphasis is placed in translated texts on the feelings a destination might evoke, particularly that of enjoyment, with *enjoy* being by far the most frequently used Emotion verb. While enjoyment certainly forms part of the tourism experience, societies that appear to have reached the post-sightseeing phase tend to focus less on it. This means that the resulting target text might be less appealing to an English-speaking readership. In terms of what can be enjoyed, the most frequent element is *view*, followed by *panorama*. Other things that can be enjoyed involve a *stroll*, a *meal*, and a *tour*. *Beauty* and *peace*, the abstract elements of a destination, are very rarely mentioned. This provides further evidence that this semantic category is not adapted to the expectations of an English-speaking readership.

Overall, English translations of French texts might at first sight appear as largely successful, because of the low degree of adaptation required. However, a closer inspection reveals that target texts stay very close to their source texts with very limited attempts towards adaptation and even some unexplained patterns (i.e. lack of experience-related verbs). Although French, American, and British societies gravitate towards the post-sightseeing phase, there are differences between them with American and British societies being further ahead on the continuum. Target texts do not appear to have taken these differences into account, perhaps because on the surface French tourism discourse might appear as being similar to English.

7.1.2 Translations into English from Greek

Adaptation might be expected in most of the semantic categories when Greek tourism texts are translated into English, as a result of the similarities, based on statistical significance tests, identified during the previous stage of analysis between English (both American and British) and Greek tourism texts, apart from Perception verbs, where no statistical significant difference has been established between the English and Greek non-translated tourism texts. Successful adaptation is, however, noted only with Discovery verbs while Motion verbs are adapted in the opposite direction, and no adaptation is observed with Emotion verbs. Adaptation is also observed with Perception verbs, even though it might not be expected. Overall, a rather inconsistent approach to adaptation is noted with no clear pattern. Figure 7.2 summarises the quantitative findings presented earlier in Table 7.2.

Discovery verbs, the only successful case of adaptation, generally represent a lower frequency in both American (21.82%) and British English (21.55%) than in Greek (31.54%) tourism texts. During translation, their frequency is indeed found to decrease (22.77%), bringing them close to English texts. Example 7.1 illustrates how adaptation looks in translation. In this example, the Discovery verb *συναντώ* (*meet*) has been replaced with the Perception verb *see*. Shifts like these bring the translated tourism text in line with how English-speaking reader-tourists are likely to perceive tourist behaviour, as encoded in non-translated tourism texts, focusing less on what can be found at a destination. However, the stronger preference towards the outcome of the exploration

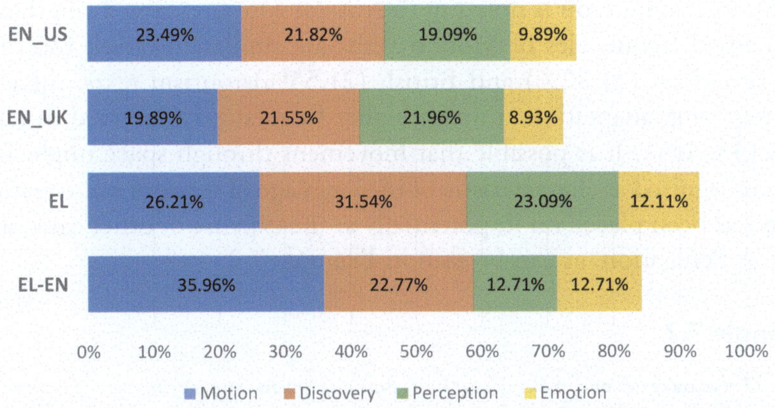

Fig. 7.2 Normalised frequencies of most prominent semantic in non-translated English, Greek source texts and translated English from Greek as % of all identified verbs

found in Greek tourism texts (with verbs such as *βρίσκω* (*find*)) is also present in their English translations, meaning that this particular element of Discovery has not been successfully adapted to focus more on open-ended exploration through verbs such as *explore*, which would be in accordance with the post-sightseeing phase of tourism Anglophone societies are currently experiencing.

Example 7.1

ST *Το θυμάρι (Corridothymus capitatus), το φασκόμηλο (Salvia futicosa), η μαλοτήρα (Sideritis syriaca spp. syriaca), η αγριομαντζουράνα (Origanum microphyllum), το δίκταμο (Origanum dictamnus) είναι ορισμένα μόνο από αυτά που μπορεί να* ***συναντήσει*** *και, κυρίως, να μυρίσει, κανείς στο νησί.*

BT *Thyme (Corridothymus capitatus), sage (Salvia futicosa), marjoram (Sideritis syriaca spp. syriaca), wild marjoram (Origanum microphyllum), dictam (Origanum dictamnus) are just some of the things one can* ***find*** *and, above all, to smell, on the island.*

TT *The thyme (Corridothymus capitatus), the sage (Salvia futicosa), the mountain tea (Sideritis syriaca spp. syriaca), the wild marjoram (Origanum microphyllum), the dittany (Origanum dictamnus) are only some of the herbs that one can* ***see*** *and smell on the island.*

Source www.incrediblecrete.gr

Additional adaptation is noted in the case of Motion verbs. Even though normalised frequencies of Motion verbs are similar in Greek (26.21%) and American (21.82%) and British (21.55%) tourism texts, therefore not requiring adaptation, a much higher frequency is observed in target texts (35.96%). It is possible that movement through space might have been recognised as characteristic of the language of tourism but too much focus has been placed on it, potentially at the expense of other categories, such as Perception, as can be seen in Example 7.2.

Example 7.2

ST *Ο πολιούχος της πόλης μαζί με την εκκλησία της Αγίας Παρασκευής, με το ηλιακό ρολόι στο προαύλιο και τον ζωδιακό κύκλο – ηλιοτρόπιο σκαλισμένο στο δάπεδο του 1911, αποτελούν τα κορυφαία προσκυνήματα του Γαλαξιδίου.*

BT *The patron saint of the city together with the church of Agia Paraskevi, with the sundial in the forecourt and the zodiac—sunflower carved on the floor from 1911, are the top pilgrimages of Galaxidi.*

TT *It is one of the prominent religious spots of Galaxidi, with the church of Agia Paraskevi being the other.* ***Visit*** *it to marvel at the sun clock in its yard and the zodiac circle drawn on its floor of 1911.*

Source www.visitgreece.gr

In this example, the reader-tourist is not being directly addressed in the Greek source text. Instead, a long description of the attractions of the destination is offered in the hope that this will create in the reader the desire to visit the destination. In the English translation some necessary restructuring occurs to address the long sentence, but also a Motion verb is introduced. It is possible, although more evidence is needed, that in cases where syntactic restructuring is required, Motion verbs are used as 'fillers', because of their importance in tourism discourse. However, the resulting text reflects the sightseeing tourism phase where more emphasis is placed on prototypical elements of tourism, such as movement, and not on the post-sightseeing one that is currently being experienced by the Anglophone readership. This can have significant implications for the success of the tourism message.

Even though not required, adaptation is observed with Perception verbs. Their distribution has been found to be similar in Greek

(23.09%), American (19.09%), and British English (21.96%) tourism texts. However, their frequency is lower in the English target texts (12.71%). One of the reasons for this reduction in the frequency of Perception verbs is the absence of the verb *experience* in translated texts. Although, the verb *ζω* (*live*) has been identified in the Greek tourism texts, albeit with a very low frequency, its equivalent is not found in the target texts, similar to what has been observed with English translations from French. In Example 7.3, the whole clause focusing on the experience is completely omitted in the target text. This is a critical omission, as societies that have reached the post-sightseeing phase, like the American and the British ones, tend to pay a lot of attention to experiences.

Example 7.3

ST *Τα νησιά του Ιονίου και του Αιγαίου και οι ακτογραμμές της ηπειρωτικής Ελλάδας σας περιμένουν, για να* ***ζήσετε*** *το ελληνικό καλοκαίρι και να χαρείτε τον ήλιο και τη θάλασσα μαζί με τους αγαπημένους σας.*

BT *The islands of the Ionian and Aegean and the coastlines of mainland Greece are waiting for you, to* ***experience*** *the Greek summer and enjoy the sun and the sea with your loved ones.*

TT *Islands throughout the Ionian and Aegean and coastal resorts on the mainland await you.*

Source www.discovergreece.com

Finally, no adaptation is observed with Emotion verbs: their frequency is almost identical in Greek source texts (12.11%) and English target texts (12.71%) and higher compared to that found in American (9.89%) and British (8.93%) tourism texts. What is typically enjoyed in the English target texts is the *view*, followed by specific things like *sports*, *activities*, or *nature*, while *beauty* is very rarely mentioned. Overall, English translations of Greek texts put more emphasis on pleasure than non-translated tourism texts in American and British English, while they also focus on more concrete aspects of the destination that can be enjoyed. This results in a concept of tourist behaviour that is not entirely in line with the post-sightseeing phase of tourism and is likely to impact the promotional potential of the text.

Overall, while a limited attempt towards adaptation is noted in the case of English translations of Greek texts, no clear pattern can be identified. Even though target texts appear to at least manage to successfully adapt Discovery verbs, the fact that some categories are adapted when not required, as well as that target texts generally tend to stay close to their Greek source texts, indicates that the resulting tourism text does not manage to capture tourist behaviour as it is perceived in societies that experience the post-sightseeing phase of tourism.

7.1.3 Translations into English from Russian

A very high degree of adaptation is expected in Russian-English translations, with all semantic categories requiring adaptation, based on the differences noted between English (both American and British) and Russian non-translated texts during the previous stage of analysis. Successful adaptation is noted only for Perception and Emotion verbs, while weak adaptation is noted for Discovery verbs. No adaptation is observed for Motion verbs. Overall, some attempt towards adaptation is noted. Figure 7.3 summarises the quantitative findings presented earlier in Table 7.2.

Successful adaptation is only noted for Emotion verbs. Verbs denoting Emotion have not been noted as frequently used during the examination of the Russian non-translated corpus component, but they are found in the English translations (6.90%), and they are distributed with a similar frequency to that found in both American (9.89%) and British (8.93%) English texts. There is a clear recognition that Emotion verbs play a role in tourism discourse in English and therefore they need to be introduced during translation. However, there is also a recognition that too much emphasis should not be put on such verbs, considering that societies that are experiencing the post-sightseeing phase seem to be moving away from them gradually. The only Emotion verb used in an English translation is *enjoy*, which is often used to replace other verbs, as can be seen in Example 7.4. Here, the Motion verb *погулять* (*walk*) has been replaced with *enjoy*, shifting the focus to sensation.

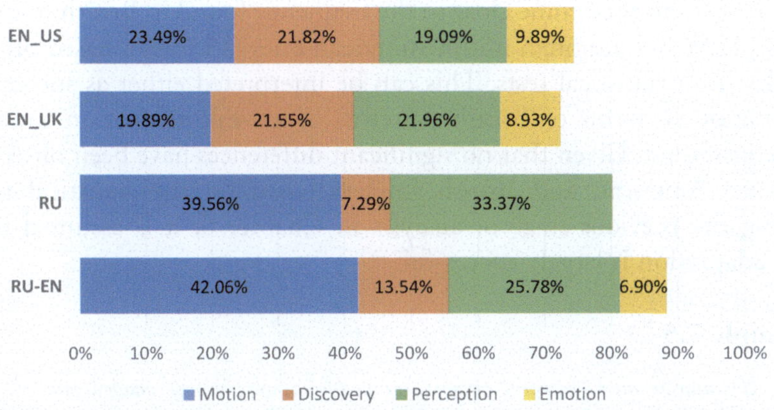

Fig. 7.3 Normalised frequencies of most prominent semantic categories in non-translated English, Russian source texts, and translated English from Russian as % of all identified verbs

Example 7.4

ST *Гости могут погулять по живописному парку со старыми деревьями, увидеть сохранившиеся исторические объекты, а также оценить новые современные вольеры с уникальными животными.*

BT *Guests can* ***take a walk*** *in the picturesque park with old trees, see the preserved historical objects, and also appreciate the new modern enclosures with unique animals.*

TT ***Enjoy*** *the beautiful park with old trees while studying extant historical artifacts. Also, appreciate the new era of one of the Kaliningrader' favorite places, the contemporary open- air cages with rare animals.*

Source www.russia.travel

Examining more closely what can be enjoyed at the destination, *views*, *vistas*, and *panoramas* are most frequently mentioned, followed by food-related items and activities. No references are made to more abstract elements of the destination, with the only exception being *untouched wilderness*, mentioned once. Therefore, the resulting target text still in some respect belongs more to the sightseeing phase.

There is also evidence of adaptation in the category of Perception, albeit weak. Verbs denoting Perception represent a lower frequency of all verbs in English target texts (25.78%) than in Russian source texts

(33.37%), and they come close to the frequency found in British tourism texts (21.96%), although not in American ones (19.09%), based on the results from statistical tests. This can be interpreted either as successful adaptation in terms of British patterns, or a weak adaptation towards American ones. Given that no significant differences have been observed between American and British English based on the statistical tests during the previous stage of analysis in Chapter 6, it is assumed here that adaptation is largely successful.

Example 7.5

ST *Несмотря на плачевное состояние усадьбы, некоторые усадебные постройки все же сохранились: сейчас посетители могут* **увидеть** *здания конного двора, мельничного флигеля и храм XIX века, а также покормить зверей в бельчатнике.*

BT *Despite the deplorable state of the manor, some of the manor buildings still survived: now visitors can* ***see*** *the buildings of the horse yard, the mill outbuilding and the temple of the nineteenth century, as well as feed the animals in the squirrel cage.*

TT *Despite the sorry state of the estate, some of the manor buildings survived: the restored horse yard, mill wing, and nineteenth century church* ***are now open*** *for visitors. Visitors also have a chance to feed the animals in the squirrel pen.*

Source www.discovermoscow.com

In Example 7.5, the Perception verb *увидеть* (*see*) where the reader-tourist is the subject has been replaced with a description of the buildings highlighting the fact that they can be visited. Such shifts bring the translated text closer to the expectations of the Anglophone reader-tourist placing less emphasis on what can be seen at a destination, which is one of the most prominent elements of Russian tourism discourse reflecting the sightseeing phase. However, the fact that the reader-tourist is not mentioned and instead a long description is used, appears to contradict the communicative conventions of English tourism texts. Furthermore, no emphasis appears to be put on experiencing the destination, since the verb *experience* is not found in the English translations of the Russian tourism texts. As with Emotion verbs, although an attempt is made towards adaptation, the resulting tourism text represents a notion of tourism behaviour that still heavily belongs to the sightseeing phase.

Another example of weak adaptation is found with verbs denoting Discovery. Although this category is prominent in both American (21.55%) and British (21.82%) tourism texts, it is considerably less so in Russian (7.29%) texts. An increase in the frequency of Discovery verbs is found in the English target texts (13.54%); however, this is not high enough to come close to English normalised frequencies. Nevertheless, there seems to be some recognition that Discovery verbs are important for the conceptualisation of tourist behaviour in English, as can be seen in Example 7.6, where the Motion verb *выйти* (*exit*) is replaced with the Discovery verb *find*.

Example 7.6

ST *Если подняться по лестнице в кафе* Surf Coffee, *то можно* **выйти** *на балкон, откуда открывается вид на Яузу, Андроников монастырь и виадук, по которому поезда подъезжают к Курскому вокзалу.*

BT *If you climb the stairs to the Surf Coffee cafe, you can* ***exit*** *to the balcony, from where you can see the Yauza, the Andronikov Monastery and the viaduct, along which trains arrive at the Kursk railway station.*

TT *If you go up the stairs to Surf Coffee, you will* ***find*** *a balcony offering a splendid view of the Yauza River, the Andronikov Monastery, and the viaduct allowing overpass to the Kursky railway terminal.*

Source www.discovermoscow.com

Motion verbs were found to be more frequently used in non-translated Russian tourism texts compared to English texts, contrary to what has been observed for Discovery verbs. Therefore, the shift here successfully manages to reconcile the differences between the two languages in terms of how tourist behaviour is conceptualised. However, more shifts of this type would have been required for adaptation to be successful in the target texts overall. Additionally, the only Discovery verb found is *find*, which suggests that attention is given only to the outcome of the exploration, without focusing at all on the process of exploring the destination. This is tightly linked with the sightseeing phase which Russian society is experiencing.

Finally, no adaptation is observed with verbs denoting Motion. Motion represents a high frequency of all verbs in both source (39.36%) and target texts (42.06%) and is far away from frequencies found in

American (23.49%) and British (19.89%) tourism texts. This is a clear case of target texts staying very close to their source texts. What is more, Motion verbs found in the English target texts (i.e. *visit*, *come*, *go*) encode movement in general, rather than specific information on how movement occurs, which is again linked to the sightseeing phase.

In sum, some degree of adaptation is observed in the English translations from Russian and there is an attempt to make the target texts encode some elements of the post-sightseeing phase of tourism to make them more attractive to an Anglophone audience. However, these attempts are not comprehensive enough and the resulting target texts reflect a concept of tourist behaviour that largely occupies a middle point in the sightseeing/post-sightseeing continuum.

Overall, the three language pairs examined here exhibit very different approaches towards adaptation. The highest degree of adaptation is observed when Russian texts are translated into English, while the lowest occurs when the source texts are French. It is possible that when differences are pronounced, there might be more scope for adaptation to occur. Once the translations into Russian are examined (Sect. 7.3), it may be possible to say whether there is any further evidence to support this hypothesis.

No clear pattern of adaptation (or lack thereof) is observed with English translations of Greek texts; no clear rationale can be detected as to why some semantic categories of verbs are adapted even though adaptation might not be expected in those cases. Regardless of the adaptation patterns in each language pair, the resulting target texts generally do not manage to encode the post-seeing tourism phase that the Anglophone world appears to be experiencing.

7.2 Tourism Texts Translated into French

The next stage of the analysis focuses on the adaptation patterns found in tourism texts translated into French. A smaller number of websites are examined here, as a result of the availability of data included in the TrAIL corpus. Specifically, although all English websites (both American and British) examined here offer a French version, only two Greek ones do, while no translations into French from Russian official tourism websites have been identified. While this affects to some extent the comparability of results, it is considered an inevitable practical limitation of a multilingual corpus design. Results are in all cases normalised and therefore comparisons can be made even across corpus components that are uneven in terms of size.

When examining translations into French, as before, the first element to be examined is the overall distribution of verbs employed in tourism texts, which meet the threshold frequency measure of 0.18ptw and which also have the reader-tourist as their subject/agent in non-translated and translated texts and their source texts, to establish the basis for the calculations of the normalised frequency of verbs belonging to specific semantic categories (Table 7.4).

Next, the distribution of the four most prominent semantic categories is examined to reveal adaptation patterns. The normalised frequencies of

Table 7.4 Normalised frequencies of most frequent verbs with reader-tourist as subject/agent in non-translated French, source texts (i.e. non-translated texts in American English, British English, Greek), and translated French corpus components

		Tokens	Verbs (types)	Verbs (tokens)	
				f	nf (ptw)
Non-translated texts	FR	170,060	27	2,018	11.86
Source texts	EN_US	170,520	35	2,750	16.12
	EN_UK	170,704	53	5,082	29.77
	EL	170,601	20	1,503	8.81
Translated texts	EN_US-FR	196,771	54	3,722	18.91
	EN_UK-FR	128,742	54	3,730	28.97
	EL-FR	130,623	13	865	6.62

the verbs that were identified in the preliminary stage of the analysis are presented in Table 7.5.

Table 7.6 offers an overview of types of adaptation observed for each language pair. Unlike translations into English examined earlier, little adaptation is generally required in the case of translations into French, because of the similarities, that is, a lack of statistically significant difference, between non-translated texts in French and the other languages examined here, that is, American English, British English, and Greek.

What is also easily noticeable from the table is that there are no instances of successful adaptation, even for the few categories where this might be required. It seems that the general pattern is for target texts to stay close to their source texts, but a closer look at how the semantic categories are realised exactly is needed to reach more valid conclusions.

7.2.1 Translations into French from American English

When examining translations from American English into French, no semantic category appears to require adaptation, because of the similarities between French and American English based on the statistical tests during the previous stage of analysis in Chapter 6, suggesting that generally tourism discourse in French and American English share many similarities, which can facilitate the translation process and make it easier to create an image of tourism that is compatible with reader-tourists' perceptions. This appears to have been recognised and indeed no adaptation is observed with any of the semantic categories. Figure 7.4 summarises the quantitative findings presented earlier in Table 7.6.

While the overall distribution of semantic categories might not require adaptation, the way these are realised through specific verb choices does, also considering that it has been found in the previous stage of analysis that the American society appears to be experiencing the post-sightseeing phase of tourism, while the French one is somewhere between the sightseeing and post-sightseeing one.

The previous stage of analysis has revealed meaningful differences regarding Discovery verbs, and specifically French non-translated texts

Table 7.5 Normalised frequencies of most prominent semantic categories of verbs in non-translated French, source texts (i.e. non-translated texts in American English, British English, and Greek), and translated French corpus components as a percentage of identified verbs (tokens)

		NT TEXTS (TL)	NT TEXTS (SL)			TRANSLATED TEXTS		
		FR	EN_US	EN_UK	EL	EN_US- FR	EN_UK- FR	EL- FR
Verbs (tokens)		2,018	2,750	5,082	1,503	3,722	3,730	865
	Raw frequency	535	646	1,011	394	983	1,225	269
Motion	Normalized frequency	26.51%	23.49%	19.89%	26.21%	26.41%	32.84%	31.10%
	Raw frequency	515	600	1,095	474	752	705	300
Discovery	Normalized frequency	25.52%	21.82%	21.55%	31.54%	20.20%	18.90%	34.68%
	Raw frequency	393	525	1,116	347	660	753	179
Perception	Normalized frequency	19.47%	19.09%	21.96%	23.09%	17.73%	20.19%	20.69%
	Raw frequency	245	272	454	182	358	304	86
Emotion	Normalized frequency	12.14%	9.89%	8.93%	12.11%	9.62%	8.15%	9.94%

Table 7.6 Types of adaptation observed in translations into French (verbs)

	EN_US-FR	EN_UK-FR	EL-FR
Not Required	Motion Discovery Perception Emotion	Discovery Perception	Motion Perception Emotion
Additional			
No		Emotion	Discovery
Successful			
Weak			
Over		Motion	
Reverse			

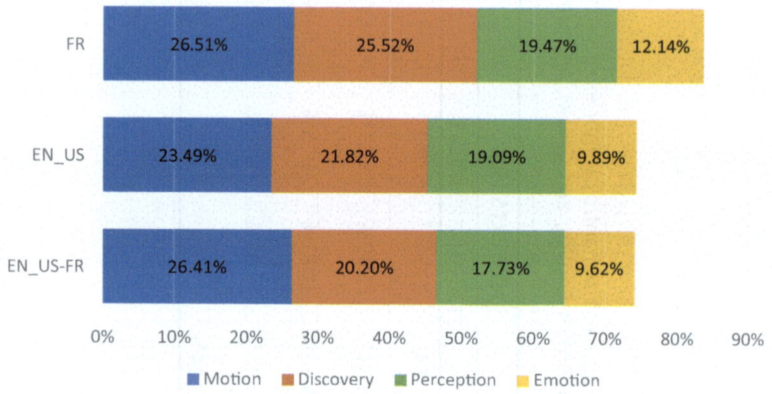

Fig. 7.4 Normalised frequencies of most prominent semantic categories in non-translated French, American English source texts, and translated French from American English as % of all identified verbs

have been found to focus less on the process of exploration and more on the outcome of it compared to American English tourism texts. Target texts appear to follow patterns found in their source texts, with significantly more attention being placed on exploration compared to what is typically found in French tourism discourse. Specifically, *explorer* (*explore*) represents 33% of all Discovery verbs in French translations, compared to 10% in French non-translated texts, and 44% in American ones. Therefore, no evidence of successful adaptation can be noted here and the resulting target texts represent a concept of tourism behaviour that corresponds more closely to the post-sightseeing phase.

Perception verbs were examined to identify the frequency of those that focus on experiencing the destination with all the senses. In the previous stage of analysis it was found that American English texts show a stronger preference towards experience-related verbs (21% of all Perception verbs) compared to French (10%). In French translations of American English texts, they represent a similarly low frequency (10%) as that found in non-translated French texts, therefore providing evidence of successful adaptation. This is illustrated in Example 7.7 where the experience-related Perception verb *experience* is replaced with the Discovery verb *découvrir* (*discover*) in a successful case of adaptation.

Example 7.7

ST *From traditional music and dance to beautiful Native art, there are many opportunities for Alaska visitors to* ***experience*** *Alaska's culture and history.*

TT *Des musiques et des danses traditionnelles à l'art amérindien dans toute sa splendeur, les occasions ne manquent pas de* ***découvrir*** *la culture et l'histoire de l'Alaska.*

BT *From traditional music and dance to Native American art in all its glory, there are plenty of opportunities to* ***discover*** *Alaskan culture and history.*

Source www.visittheusa.com

Finally, the elements that can be enjoyed at a destination were examined in relation to Emotion verbs. In the French target texts, the most frequently mentioned element is *vue* (*view*), followed by *soleil* (*sun*), *activités* (*activities*), *spectacle* (*show*), and *beauté* (*beauty*). The focus on abstract elements such as *beauty* is something that seems to characterise the post-sightseeing phase experienced by the American society and was not observed in the French non-translated texts, indicating that the French society might have not reached that stage yet. The target texts appear to have stayed very close to their source texts.

When existing similarities between American English and French non-translated tourism texts are examined, the French translations from American English succeed in creating an image of tourism that is compatible with the expectations and worldview of French reader-tourists by not adapting semantic categories that do not require adaptation. However, when these categories are examined in more detail, only a few instances of successful adaptation where the target text has been changed to appeal to the target audience have been observed. The resulting target text is therefore much closer to the concept of tourist behaviour as perceived in the French society, even though it cannot be ignored that certain aspects of it will also be appealing to an American audience.

7.2.2 Translations into French from British English

Regarding translations from British English into French, a generally medium degree of adaptation might be expected based on what has been observed during the previous stage of analysis. Adaptation is not required with Perception and Discovery verbs, and indeed it is not observed. Emotion verbs are not adapted, while adaptation in the opposite direction is noted in the case of Motion verbs. Overall, a clear pattern towards adaptation is not observed. Figure 7.5 summarises the quantitative findings presented earlier in Table 7.6.

Recalling that Motion verbs, one of the most important semantic categories in tourism discourse, are characteristic of all corpus components, it is perhaps their importance that resulted in over adaptation during translation from British English into French. The frequency of these verbs has been found to be higher in French (26.51%) than British English (19.89%) non-translated texts. An increase in their frequency (32.84%) is however noted during translation into French, suggesting that translators might have recognised the importance of this semantic category, but have overemphasised it. This excessive focus on Motion is illustrated in Example 7.8.

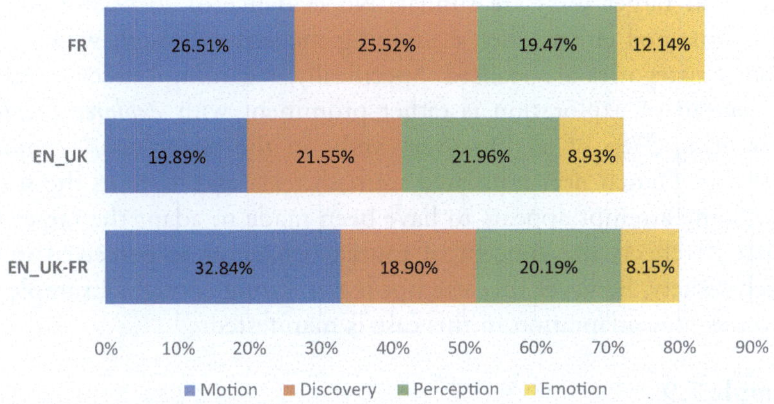

Fig. 7.5 Normalised frequencies of most prominent semantic categories in non-translated French, British English source texts, and translated French from British English as % of all identified verbs

Example 7.8

ST *You can spend an afternoon at The Beatles Story museum,* ***see*** *the houses where John Lennon and Paul McCartney grew up and enjoy a night out at the Cavern Club, where the band famously performed. Or explore some of Liverpool's 7 free museums.*

TT *Passez un après-midi au musée The Beatles Story,* ***promenez****-vous devant les maisons où John Lennon et Paul McCartney ont grandi, et découvrez les 7 musées gratuits de Liverpool.*

BT *Spend an afternoon at The Beatles Story museum,* ***stroll*** *past the homes where John Lennon and Paul McCartney grew up, and discover Liverpool's 7 free museums.*

Source www.visitbritain.com

In this example, the Perception verb *see* in English is replaced with the Motion verb *promener* (*stroll*) in the French translation. There is a clear shift here away from the senses to movement. A translation where a Perception verb, such as *admirer* (*admire*), is used would have worked equally well, suggesting that this shift is not dictated by morphosyntactic differences between the two languages.

For the remaining semantic categories, no adaptation is observed, even though it is required for Emotion verbs, and the frequencies between source and target texts are similar. No evidence of successful adaptation is obtained either when examining the verbs through which these semantic categories are realised. Specifically, regarding Discovery verbs, the element of exploration is rather prominent with *explorer* (*explore*) representing 27% of all Discovery verbs in the target texts compared to 10% in French non-translated tourism texts and 46% in the source texts. Some attempt appears to have been made to adapt the target text so that it reflects the concept of tourist behaviour as perceived in the French society, however this attempt is not strong enough. Example 7.9 illustrates how adaptation in this case is manifested.

Example 7.9

ST *Spend a day **exploring** Fort William, the Outdoor Capital of the UK and one of the most frequently used film locations in Scotland.*
TT *Passez la journée à Fort William, la capitale du plein air du Royaume-Uni et l'un des lieux de tournage les plus connus d'Écosse.*
BT *Spend the day in Fort William, the outdoor capital of the UK and one of Scotland's best-known film locations.*
Source www.visitscotland.com

The Discovery verb *explore* found in the British English source text is omitted in the French translation resulting in two sentences which, although similar in content, communicate a very different notion of tourism. The English text promotes the idea of an active tourism, where exploration is central to how the day is spent. The French translation, on the other hand, promotes a more passive idea of tourism, where tourists do not need to exert much effort, also evoking a sense of pleasure that is to be gained by simply being at the destination. This is considered an example of successful adaptation and the creation of a tourism message that is likely to be appealing to the French-speaking audience. More shifts of this type would however have been necessary to recreate the concept of tourist behaviour as perceived in the French society.

A rather odd pattern is observed with Perception verbs and specifically with experience-related ones. Experience-related verbs are employed occasionally in French (10% of all Perception verbs) and British (14%) non-translated tourism texts, but they are found rarely in French translations from British English (5%). In fact, *vivre* (*live*), the only frequently used experience-related verb found in the French non-translated component, is not observed in the translations from British English.

Example 7.10

ST *With the right safety precautions, cycling is a unique way to* ***experience*** *and enjoy the city.*
TT *Lorsque toutes les précautions de sécurité sont respectées, le vélo est un moyen exceptionnel de profiter de la ville.*
BT *When all safety precautions are followed, cycling is an exceptional way to enjoy the city.*
Source www.visitscotland.com

Example 7.10 illustrates how experience-related verbs might be adapted during translation through omission. Other strategies involve replacing *experience* with Emotion verbs, mainly *profiter* (*enjoy*) or Discovery verbs such as *découvrir* (*discover*). The range of different strategies suggests that translators are somehow aware that experience-based tourism is likely not to be appealing to French-speaking reader-tourists.

Finally, a lack of adaptation is noted with Emotion verbs and frequencies of these are similar between source (8.93%) and target (8.15%) texts, even though they are higher in French non-translated texts (12.14%). As a result, less emphasis appears to be placed on pleasure than what might be required based on the communicative conventions of French tourism discourse. This creates a different notion of tourism experience, where pleasure does not play such a key role, reflecting potentially a different phase of tourism. Additionally, the most typical things that can be enjoyed at a destination according to the French target texts are *vue* (*view*) and *atmosphère* (*atmosphere*) suggesting that attention is being paid to both concrete and more abstract elements of the destination. Once again, this appears to be a reflection of how tourist behaviour is conceptualised in the British society rather than in the French one.

In sum, there is no evidence that successful adaptation has taken place during translation from British English into French. Target texts appear to stay close to their source texts and the concept of tourist behaviour promoted, and tourism more generally, is one that corresponds to the post-sightseeing phase which the French society does not appear to have reached yet.

7.2.3 Translations into French from Greek

A low degree of adaptation might be expected in the case of French translations from Greek, as a result of the many similarities between French and Greek non-translated tourism texts identified during the previous stage of analysis, with both societies occupying more or less the same space on the sightseeing/post-sightseeing continuum. Regarding the distribution of the semantic categories, adaptation is not required with Motion, Perception, and Emotion verbs, and indeed is not observed. The only instance where frequency distribution might be expected to be adapted is Discovery verbs, where a significant difference has been identified between English (both varieties) and Greek non-translated texts, but no adaptation is observed. Overall, French target texts have been found to stay close to their Greek source texts. Figure 7.6 summarises the quantitative findings presented earlier in Table 7.6.

The frequency of Discovery verbs in French translations is close (34.68%) to that found in the Greek source texts (31.54%), even though French tourism texts employ such verbs less frequently (25.52%). This suggests that more emphasis is placed on exploration and discovery than what French-speaking reader-tourists might be accustomed to. In terms

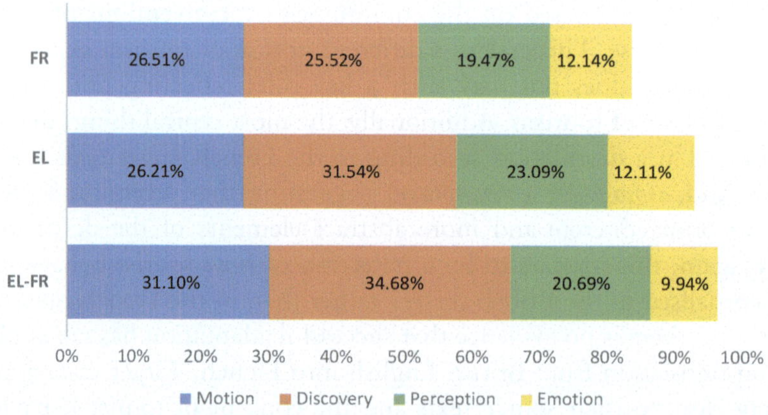

Fig. 7.6 Normalised frequencies of most prominent semantic categories non-translated French, translated French from Greek, and Greek source texts as % of all identified verbs

of the specific verbs that realise this category, Greek and French tourism discourse has been found to be very similar regarding their focus on the outcome of the exploration. However, it is possible to obtain further evidence of the lack of adaptation by examining the verb *συναντώ* (*encounter*). Greek tourism texts make use of this verb to also capture the sense of discovering something. This is the only non-translated corpus component where this is observed, reflecting some interesting differences as to how languages conceptualise semantic categories. When these texts are translated into French, the French equivalent of the verb, i.e. *rencontrer*, is employed in the translations, as can be seen in Example 7.11.

Example 7.11

ST Οι καταβόθρες είναι ένα ακόμα μοναδικό γεωλογικό φαινόμενο που θα **συναντήσετε** στην Κεφαλονιά.
BT *The katavothres (sinkholes) are another unique geological phenomenon that you will* ***encounter*** *in Kefalonia.*
TT *Les Katavothrès sont encore un autre phénomène géologique unique que vous* ***rencontrerez*** *à Céphalonie.*
BT *The Katavothres are yet another unique geological phenomenon that you will* ***encounter*** *in Kefalonia.*
Source www.discovergreece.com

A literal translation is provided in this example—an alternative option would be the use of *découvrir* (*discover*). Although the use of this alternative verb would mean that this semantic category is still not adapted, at least a potentially more explicit message to the French-speaking reader-tourist would have been conveyed. It is interesting that the use of the verb *encounter* has not been observed in English translations from Greek, further supporting the idea that there are other options available, which would better reflect the expectations of the French-speaking audience. Such literal translation is likely to affect how the tourism message is received by readers and the extent to which it corresponds to their concept of tourism.

More evidence regarding adaptation can be obtained by further examining the verb choices through which the semantic categories are realised. Because of the many similarities between French and Greek tourism discourse identified in the previous stage of analysis, the focus here is only on Motion verbs. Specifically, Greek tourism texts have been found to favour more generic Motion verbs compared to French. The French target texts have been found to stay close to their source texts, with only *marcher* (*walk*) and *promener* (*walk*) encoding more information regarding the type of movement, equivalents of which can also be found in the Greek source text. Other options which characterise the French tourism discourse such as *flâner* (*stroll*) or *arpenter* (*survey*) have not been found in French translations from Greek.

Overall, from the point of view of existing similarities, French translations from Greek appear to create an image of tourist behaviour that is largely compatible with the expectations and worldview of French-speaking reader-tourists. However, where differences between French and Greek non-translated texts are examined, it is evident that no successful attempt towards adaptation has been made and that actually the target texts stay close to their source texts. The resulting target text is closer to how tourist behaviour is conceptualised in the Greek society rather than the French, although because of the many similarities between the two, it is likely to also be appealing to a large extent to a French-speaking audience.

Generally, when tourism texts are translated into French, there are few instances where adaptation might be expected, which is most probably a result of the fact that the French society is somewhere in-between the post-sightseeing/sightseeing continuum, and therefore the French tourism discourse shares many characteristics with both the Anglophone (American and British) and Greek one. Nevertheless, even when adaptation might be expected, there is generally limited evidence that this takes place successfully, which suggests that although the resulting target text might achieve its promotional function, the representation of tourist behaviour could have been more in line with what the French-speaking audience is accustomed to.

7.3 Tourism Texts Translated into Russian

Translations from Russian are the only category where available translations from all other languages, i.e. British English, American English, French, and Greek, can be examined. This allows for more comparisons to be made and for more supporting evidence to be obtained regarding adaptation during translation across different language pairs. However, some of the corpus components are rather small due to the limited availability of data which needs to be considered when interpreting results. The starting point is once again the number of verbs employed which meets the threshold frequency measure of 0.18ptw and which also has the reader-tourist as their subject/agent in non-translated texts and in translated texts and their source texts, to establish the basis for the calculations of the normalised frequencies of verbs (Table 7.7).

Adaptation patterns are examined next by focusing on the distribution of the four most prominent semantic categories. Table 7.8 presents the normalised frequencies out of all verbs identified in each corpus component.

Table 7.9 offers an overview of types of adaptation observed for each language pair. As can be seen, adaptation is required for all semantic categories across all language pairs, that is, there is no category in the 'Not

Table 7.7 Normalised frequencies of most frequent verbs with reader-tourist as subject/agent in non-translated Russian, source texts (i.e. non-translated texts in American English, British English, French, Greek), and translated Russian corpus components

		Tokens	Verbs (types)	Verbs (tokens)	
				f	nf (ptw)
Non-translated texts	RU	170,881	14	809	4.73
Source texts	EN_US	170,520	35	2,750	16.12
	EN_UK	170,704	53	5,082	29.77
	FR	170,060	27	2,018	11.86
	EL	170,601	20	1,503	8.81
Translated texts	EN_US RU	53,430	54	1,270	23.76
	EN_UK RU	43,707	46	1,030	23.56
	FR-RU	36,464	41	796	21.83
	EL-RU	88,461	21	909	10.27

Table 7.8 Normalised frequencies of most prominent semantic categories of verbs in non-translated Russian, source texts (i.e. non-translated texts in American English, British English, French, Greek), and translated Russian corpus components as a percentage of identified verbs (tokens)

		NT TEXTS (TL)	NT TEXTS (SL)				TRANSLATED TEXTS			
		RU	EN_US	EN_UK	FR	EL	EN_US-RU	EN_UK-RU	FR-RU	EL-RU
Verbs (tokens)		809	2,750	5,082	2,018	1,503	1,282	1,030	796	909
	Raw frequency	320	646	1,011	535	394	263	354	278	326
Motion	Normalized frequency	39.56%	23.49%	19.89%	26.51%	26.21%	20.71%	34.37%	34.92%	35.86%
	Raw frequency	59	600	1,095	515	474	196	90	91	146
Discovery	Normalized frequency	7.29%	21.82%	21.55%	25.52%	31.54%	15.43%	8.74%	11.43%	16.06%
	Raw frequency	270	525	1,116	393	347	262	254	223	294
Perception	Normalized frequency	33.37%	19.09%	21.96%	19.47%	23.09%	20.63%	24.66%	28.02%	32.34%
	Raw frequency	0	272	454	245	182	103	124	51	59
Emotion	Normalized frequency	0%	9.89%	8.93%	12.14%	12.11%	8.11%	12.04%	6.41%	6.49%

Required' row, which reflects that Russian tourism texts clearly stand out from similar texts in the other languages examined here—evidence that the Russian society is still experiencing the sightseeing phase. This in turn means that there are more opportunities for adaptation and indeed more attempts towards adaptation are observed compared to other language pairs.

Finally, it is also interesting to note that no cases of additional, over or reverse adaptation have been noted; all instances belong either to the no adaptation or to successful/weak adaptation patterns. This suggests that translators either stay close to the source text or try to introduce changes that are in line with the idea of tourist behaviour as conceptualised in the Russian society.

7.3.1 Translations into Russian from American English

Although adaptation might be expected for all semantic categories when American English tourism texts are translated into Russian, based on the results obtained from the previous stage of analysis and particularly the statistically significant differences between Russian tourism texts and those in the other language examined here, only a weak attempt towards it is observed with Discovery verbs, with the distribution of the remaining categories not adapted during translation. Overall, although a high degree of adaptation is expected, there is little evidence that adaptation has occurred. Figure 7.7 summarises the quantitative findings presented earlier in Table 7.8.

As indicated, some attempt towards adaptation is noted with Discovery verbs. These verbs are used more frequently in the American English tourism texts (21.82%) than in the Russian tourism texts (7.29%). While an attempt is made to lower their frequency in the Russian target texts (15.43%), this attempt is weak and does not come close to the patterns found in the Russian tourism discourse, with frequencies standing almost exactly between the Russian and American English non-translated texts. Example 7.12 illustrates how this weak adaptation is manifested in a target text.

Table 7.9 Types of adaptation observed in Russian translations (verbs)

	EN_US-RU	EN_UK-RU	FR-RU	EL-RU
Not Required				
Additional				
No	Motion Perception Emotion	Perception Emotion		
Successful		Motion Discovery	Motion Discovery Perception	Motion Perception
Weak	Discovery		Emotion	Discovery Emotion
Over				
Reverse				

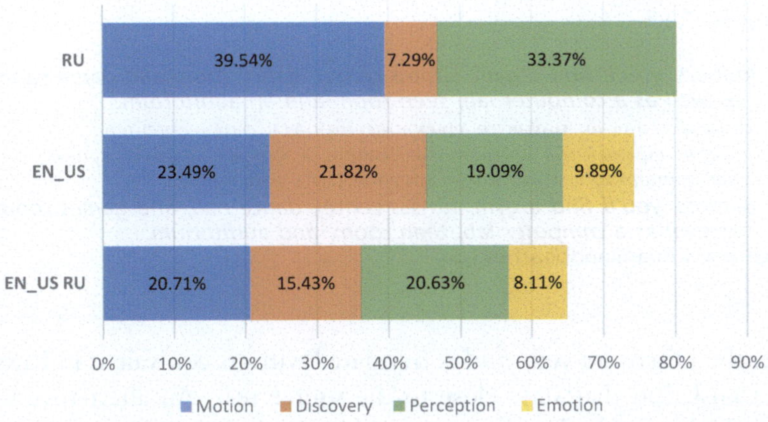

Fig. 7.7 Normalised frequencies of most prominent semantic categories in non-translated Russian, translated Russian from American English, and American English source texts as % of all identified verbs

Example 7.12

ST *Aventura is also just minutes away from Sunny Isles Beach, making this neighborhood the perfect place to **explore**, indulge, and unwind.*

TT *Aventura также всего в нескольких минутах от Sunny Isles Beach, что делает этот район идеальным местом для* **изучения,** *развлечения и отдыха.*

BT *Aventura is also just minutes from Sunny Isles Beach, making the area the perfect place for exploration, fun and relaxation.*

Source www.miamiandbeaches.com

In this example, the Discovery verb *explore* is replaced with the noun *изучения* (*exploration*). It is mostly in cases of morphosyntactic shifts like this one that the target text moves away from the source text. Therefore, the main shift here is that the reader-tourist is completely omitted, which is in line with the communicative conventions of Russian tourism texts. However, the focus on exploration is still present.

In the vast majority of cases where the verb phrase is retained in translation, a literal translation approach is followed with a verb that belongs to the same semantic category as in the source text (Example 7.13).

Example 7.13

ST *Indoors, you'll* ***find*** *a gym, fitness center, dance room and game room, as well as a computer lab, teen room and an auditorium.*

TT *В помещении вы* **найдете** *тренажерный зал, фитнес-центр, танцевальный зал и игровую комнату, а также компьютерную лабораторию, комнату для подростков и аудиторию.*

BT *Indoors, you'll* ***find*** *a gym, fitness center, dance hall, and games room, as well as a computer lab, teen room, and auditorium.*

Source www.miamiandbeaches.com

Here, the Discovery verb *find* is translated with its equivalent in Russian in a translation that stays close to the source text. An alternative solution, which would take into account the communicative conventions of tourism discourse in Russian, would be to employ the verb *есть* (*to be*) and omit the direct reference to the reader-tourist, i.e. *В помещении есть тренажерный зал … (Indoors, there is a gym…)*.

No adaptation is observed in three semantic categories, namely Motion, Perception, and Emotion. Motion verbs in Russian translations (20.71%) are not adapted to match the high frequency with which such verbs are used in Russian non-translated tourism texts (39.54%), and they remain close to the frequency found in their American English source texts (23.49%). This is further manifested by the fact that Motion is conceptualised through verbs denoting specific movement such as *прогуливаться* (*stroll*), which corresponds to the post-sightseeing phase of tourism that the American society is currently experiencing.

A similar pattern is noted in the case of Perception verbs. Although they represent a high frequency in Russian non-translated tourism texts (33.37%), the frequency found in the target texts (20.63%) reflects that found in the source texts (19.09%). Frequent references are also found in experience-related verbs such as *испытывать* (*experience*) and *погружаться* (*immerse*) which appear to be absent from Russian non-translated texts, as they reflect a post-sightseeing phase of tourism.

As far as the focus on two of the most central elements of tourism is concerned, movement through space and sight, which is highly emphasised in the Russian tourism discourse, the translated texts have not succeeded in highlighting them so that they represent a notion of tourism that is compatible with the worldview as encoded in Russian non-translated texts. At the same time, the way in which these semantic categories have been realised in the translated texts, that is through specific verb choices, appears to contradict the sightseeing tourism phase that the Russian society is currently experiencing.

Finally, Emotion verbs have not been noted in Russian tourism texts, but they are found in American English texts (9.89%). During translation, no attempt towards adaptation has been made and these verbs represent similar frequencies in target texts (8.11%). Emphasis, albeit weak, is placed on enjoying aspects of a destination, which is however something that appears not to be a prominent element of the tourism experience in the Russian society. What is more, what can be enjoyed includes both *вид* (*view*) and *красота* (*beauty*), that is both concrete and abstract elements, suggesting a clear influence from the American English source texts.

Overall, even though a high degree of adaptation is required, no clear attempt towards adaptation is observed in the case of the Russian translations from American English, with most semantic categories staying close to equivalent ones found in their source texts. Even when a weak attempt towards adaptation is noted, this is too weak to provide any convincing evidence that communicative conventions of Russian tourism texts have been considered. However, it must be noted that translations from just one American tourism website are included in this Russian component, and therefore any generalisations should be made with caution. The limited evidence obtained here suggests that the resulting tourism texts represent a post-sightseeing notion of tourism as perceived in the American society and is distant from the sightseeing one circulating in the Russian society. This might simply reflect a lack of appreciation of the need to adapt the tourism message for a new audience or a reflection of the status of American culture.

7.3.2 Translations into Russian from British English

The Russian translations from British English are split in their approach between staying close to their source texts for two of the four semantic categories examined (i.e. Perception and Emotion) and producing successful adaptations for the remaining two (i.e. Motion and Discovery). Overall, although a high degree of adaptation is expected, based on the results obtained from the previous stage of analysis and particularly the statistically significant differences between Russian and British English tourism texts, mixed evidence of adaptation is obtained. Figure 7.8 summarises the quantitative findings presented earlier in Table 7.8.

Motion verbs, which have been found to be successfully adapted, are strongly favoured in Russian tourism discourse (39.54%), and although they are also prominent in British English texts, their frequency is lower (19.89%) in the latter. However, Russian translations from British English employ a higher number of Motion verbs (34.37%) than their source texts, which is close to the frequency found in Russian non-translated texts. A reverse pattern is observed for Discovery verbs which

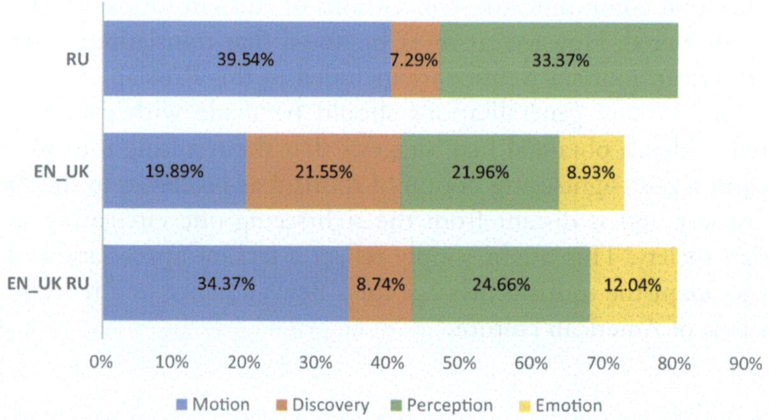

Fig. 7.8 Normalised frequencies of most prominent semantic categories in non-translated Russian, British English source texts, and translated Russian from British English as % of all identified verbs

are used less frequently in Russian translations (8.74%) than in British English source texts (21.55%), close to the frequencies found in Russian non-translated tourism texts (7.29%). For these two categories, translators have managed to successfully highlight the most prototypical aspect of tourism, that is movement through space, and downplay exploration.

Example 7.14

ST *Don't **miss** the award-winning Canterbury Tales attraction, where the sights, sounds and smells of Chaucer's medieval England are recreated.*

TT *Всем, кто хочет увидеть, услышать и почувствовать Англию такой, какой её знали герои знаменитого средневекового произведения, стоит* **отправиться** *на экскурсию Canterbury Tales Attraction.*

BT *Anyone who wants to see, hear and feel England as the heroes of the famous medieval work knew it, should **go** on the Canterbury Tales Attraction tour.*

Source www.visitbritain.com

In Example 7.14, the Action verb *miss* is replaced with the Motion verb *отправиться* (*go*), also noting a shift from a negative to an affirmative clause. This is a typical shift in Russian translations from British English found frequently in the data. Overall, the preference for *отправиться* (*go*) is so strong in Russian translations of British English texts, that this verb is employed approximately three times more frequently compared to *go* in the source texts. This also demonstrates an attempt to use more verbs related to non-specific movement in Russian translations in line with what has been found in Russian non-translated texts. However, specific verbs such as *прогуливаться* (*stroll*) are also used suggesting that there is potentially more scope for adaptation.

Example 7.15 illustrates how Discovery verbs have been adapted during translation. *Explore*, a frequently used Discovery verb in British English, is replaced by *ознакомиться* (*get to know*) which, although similar in meaning, places much more emphasis on the cognitive outcome of the encounter and is therefore classified as a Cognition verb.

Example 7.15

ST *A great way to **explore** the city while staying active, best start practicing your downward dog and sun salutations now!*
TT *Так что если вы хотите* **ознакомиться** *с городом и одновременно получить пользу для здоровья, будьте готовы принять классические позы "собака мордой вниз" и "приветствие солнцу"!*
BT *So if you want to **get to know** the city and get some health benefits at the same time, be prepared to do the classic down-facing dog and sun salutation poses!*
Source www.visitbritain.com

Nevertheless, the verb *исследовать* (*explore*) is also occasionally used in Russian target texts, and although its frequency is rather low, contrary to what happens in the source texts, an element of the post-sightseeing phase of tourism appears to be introduced in the target texts.

No adaptation is observed in the case of Perception and Emotion verbs, even though adaptation would have been required, as a result of the similarities between Russian and British English non-translated texts regarding these semantic categories. In both cases, Russian target texts follow closely the patterns and frequencies found in British English source texts. As a result, Perception verbs are used less frequently in translated Russian texts (24.66%) than non-translated texts (33.37%), close to what is found in source texts (21.96%). Further evidence of the lack of adaptation of Perception verbs is obtained from the occasional use of the experience-related verb *окунуться* (*immerse*) in the target texts, one that is not found as frequently in non-translated Russian tourism texts.

Conversely, Emotion verbs have not been found in Russian non-translated tourism texts but they are noted in translated texts from British English (12.04%) reflecting what is found in the source texts (8.93%). This results in emphasis being paid to enjoying aspects of the tourism experience in Russian tourism discourse, potentially at the expense of another prototypical element of tourism, that is, Perception. Even though both semantic categories encode mental states, there is a fundamental difference between, for example, looking at a sunset and enjoying it. Hedonism and the pursuit of pleasure is different from sightseeing and it has been argued in the previous stage of analysis that it characterises mainly societies that are in the middle point of the continuum—it is a phase that societies need to go through before reaching the post-sightseeing one. Therefore, emphasis on sensations, and pleasure in particular, through verbs such as *наслаждаться* (*enjoy*)—by far the most frequently used Emotion verb—position the resulting tourism discourse further away from the sightseeing phase.

It is worth noting that Emotion verbs might be used even when no such verb is used in the source text, as in Example 7.16. In this example, the verb *наслаждаться* (*enjoy*) is added in the target in an attempt to unpack the phrase *relax amid the chilled out atmosphere*. Although such instances of addition are not frequent, they go against the direction of adaptation that would have been required here.

Example 7.16

ST *Relax amid the chilled-out atmosphere of Princes Street Gardens in the shadow of the majestic Edinburgh Castle, with its famous floral clock and manicured grounds.*

TT *Вдохните свежий воздух и* **насладитесь** *умиротворяющей атмосферой Садов Принцесс-стрит под стенами величественного Эдинбургского замка.*

BT *Breathe in the fresh air and* ***enjoy*** *the peaceful atmosphere of Princess Street Gardens under the walls of the majestic Edinburgh Castle.*

Source www.visitbritain.com

Overall, while a high degree of adaptation might be expected in the case of British English translations into Russian, target tourism texts fluctuate between attempts towards adaptation which are successful, and staying close to the source texts. The resulting target text largely reflects a post-sightseeing phase of tourism, which might not be appealing to the Russian-speaking readership.

7.3.3 Translations into Russian from French

Russian translations from French make a more successful attempt towards adaptation compared to translations from English. Successful adaptation is noted in the case of Motion, Discovery, and Perception verbs, and also a weak attempt towards adaptation is observed in the case of Emotion verbs. Overall, there is evidence of a high degree of adaptation. Figure 7.9 summarises the quantitative findings presented earlier in Table 7.8.

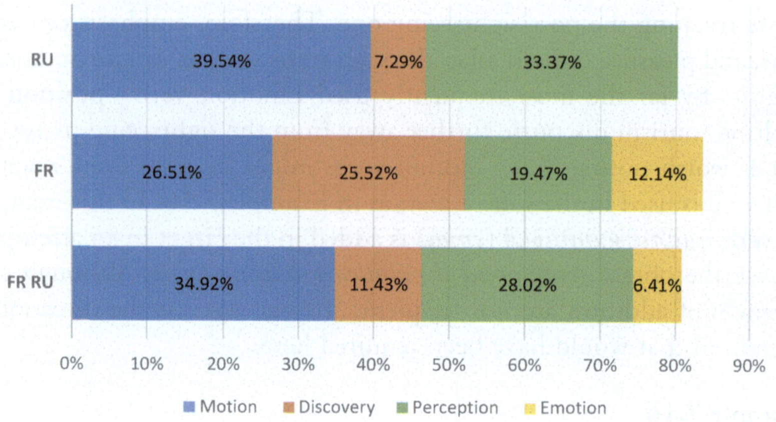

Fig. 7.9 Normalised frequencies of most prominent semantic categories in non-translated Russian, French source texts, and translated Russian from French as % of all identified verbs

Russian translations from French employ a higher frequency (34.92%) of Motion verbs than their French source texts (26.51%), close to the frequency found in Russian non-translated tourism texts (39.54%), resulting in successful adaptation. The importance of movement through space in Russian tourism texts appears to have been recognised, in line with the sightseeing phase of tourism that the Russian society is currently experiencing.

Example 7.17

ST *Après la culture, la nature: on va se balader au parc de la Gloriette, un poumon vert de près de 200 hectares au sud de la ville.*

BT *After culture, nature: we go for a walk in the Parc de la Gloriette, a green lung of nearly 200 hectares to the south of the city.*

TT *Выполнив культурную программу, можно* **отправиться** *на природу: прогуляйтесь в южной части города в парке Глориетт площадью около 200 гектаров.*

BT *After completing the cultural program, you can* ***go*** *to nature: take a walk in the southern part of the city in Gloriette Park with an area of about 200 hectares.*

Source www.france.fr

In Example 7.17, the Russian Motion verb *отправиться* (*go*) is introduced as part of a fairly dramatic syntactic reorganisation. In the French source text, there are no verbs in the elliptical first part of the sentence; there are a number of options as to what the omitted verb might be, including an option where the reader-tourist is directly addressed. It is this interpretation that the translators have opted for and therefore have added a Motion verb, which appears to have worked well as it corresponds with the notion of tourism in the Russian society.

However, the choice of specific verbs to realise Motion suggests that some influence from the French source texts might be exerted. This is mainly as a result of the use of the verb *прогуливаться* (*stroll*) found in translated texts, which has not been noted as frequently found in Russian non-translated texts.

Successful adaptation is also noted in the case of Discovery verbs. These are used less frequently in Russian translations (11.43%) than in their French source texts (25.52%), closer to what is found in Russian tourism texts (7.29%). Therefore, less emphasis is put on exploration, in line with the notion of tourist behaviour as reflected in non-translated Russian tourism texts.

Example 7.18

ST *En toile de fond, le plus long fleuve de France, la Loire qui dessert une vingtaine de châteaux à visiter et autant d'histoires à redécouvrir, de jardins à humer, de forêts à* ***explorer****, de vins à goûter...*

BT *In the background, the longest river in France, the Loire which serves around twenty castles to visit and as many stories to rediscover, gardens to smell, forests to* ***explore****, wines to taste...*

TT *На берегах пересекающей регион Луары, самой длинной реки во Франции, расположено около 20 замков, каждый со своей уникальной историей. Здесь же раскинулись цветущие сады,* **таинственные леса** *и винодельческие поместья, где можно продегустировать местные вина...*

BT *On the banks of the Loire region, the longest river in France, there are about 20 castles, each with its own unique history. Blooming gardens,* ***mysterious forests*** *and wine estates are also spread here, where you can taste local wines...*

Source www.france.fr

Example 7.18 is another instance of fairly dramatic syntactic reorganisation but this time involving a Discovery verb. The French phrase *forêts à explorer* (*forests to explore*) creates an image of forests waiting for the tourists to come and do the exploration. Although the subject is omitted here, it is clear that it is the reader-tourist. The Russian translation completely restructures the clause and replaces the verb with the noun phrase *таинственные леса* (*mysterious forests*) omitting how one is to interact with these, indirectly luring the reader-tourist to explore them by describing them as *mysterious*. Descriptions appear to be powerful means of destination promotion in Russian tourism discourse. The rhetorical effect achieved by the parallel construction aims to create a lasting effect on the reader creating a desire in the reader to visit the destination. As with Motion verbs however, there is evidence of influence from French source texts regarding the choice of verbs. The most frequent Discovery verb in the target texts is *исследовать* (*explore*) suggesting a clear emphasis which has been found to be absent from non-translated texts. This introduces elements of the post-sightseeing phase of tourism which are likely to be at best alien to the target readership.

The final semantic category where successful adaptation has been noted is Perception. Russian non-translated tourism texts show a strong preference towards these verbs (33.37%) and their frequency is higher in target texts (28.02%) than in their source texts (19.47%). As with Motion, the importance of this semantic category appears to have been recognised and more emphasis is put on it during translation. Example 7.19 illustrates what adaptation looks like with Perception verbs.

Example 7.19

ST ***Découvrir*** *les impressions de Monet à Giverny.*
BT ***Discover*** *Monet's impressions in Giverny.*
TT **Увидеть** *Живерни глазами Моне.*
BT ***See*** *Giverny through Monet's eyes.*
Source www.france.fr

In this example, the Discovery verb *découvrir* (*discover*) is replaced with a Perception verb, specifically one related to sight, *увидеть* (*see*), during translation. This is a good example of how the frequency of Discovery verbs is reduced during translation, while that of Perception verbs is increased. The focus on sight is further highlighted through the use of the noun *глаза* (*eyes*)—there is a very clear emphasis on sight in this example, which is representative of what successful adaptation looks like. Nevertheless, once again some evidence of influence from the source texts might be found in the use of the experience-based verb *погружаться* (*immerse*) in translated texts which reflects a post-sightseeing phase of tourism.

Finally, weak adaptation is noted in the case of Emotion verbs. These have not been found as frequently used in Russian tourism discourse, but they are found in French tourism texts (12.14%). These are also found in Russian translations from French and with a lower frequency (6.41%) suggesting that some attempt towards adaptation has been made. Enjoying things at a destination or focusing on emotions is not elements of tourism that are considered important in Russian tourism discourse, yet they are found in translations from French. Therefore, a message, such as that found in Example 7.20, is in contrast to the sightseeing phase of tourism that the Russian society is currently experiencing.

Example 7.20

ST *Et au sommet, on* ***apprécie*** *le panorama à 360° sur l'ensemble de l'archipel, les fumerolles et vapeurs de soufre en prime !*

BT *And at the top, we* ***appreciate*** *the 360° panorama over the whole archipelago, the fumaroles and sulfur vapors as a bonus!*

TT *На вершине* **оценим** *круговой панорамный вид на весь архипелаг, а также фумаролы и сернистые испарения!*

BT *At the top we will appreciate a 360-degree panoramic view of the entire archipelago, as well as fumaroles and sulfurous fumes!*

Source www.france.fr

Here, the reader is told exactly what they will experience in terms of emotions in a clause that also uses the first-person plural pronoun in Russian to create a sense of rapport between the writer and the audience. However, such literal renditions are generally rare in Russian translations from French. More frequently, even though Emotion verbs might not be characteristic of Russian tourism discourse, they are added during translation as seen in Example 7.21.

Example 7.21

ST *Après avoir visiter Bordeaux, on peut aussi filer plein sud vers le bassin d'Arcachon pour* ***tâter l'eau de l'océan.***

BT *After visiting Bordeaux, you can also head south towards the Arcachon basin to* ***test the waters of the ocean.***

TT *Посетив Бордо, вы можете отправиться на юг в сторону лагуны Аркашон, чтобы* **почувствовать дыхание океана.**

BT *After visiting Bordeaux, you can head south towards the Arcachon lagoon to* ***feel the breath of the ocean.***

Source www.france.fr

The example above clearly demonstrates the extent to which the target text might be adapted. The French source text describes swimming as an activity that tourists can do at the destination, using the expression *tâter l'eau de l'océan* (*test the ocean water*) both with a literal and metaphorical meaning. The focus on the activity is completely shifted in the Russian target text and emphasis is now placed on the sensations that the ocean can offer, specifically on the sounds of the waves. Why such a shift might be required is not clear, but this example offers convincing evidence that adaptation is extensive in the French-Russian tourism translation.

Overall, a relatively high degree of adaptation is noted in the Russian translations from French especially regarding the overall distribution of semantic categories. Even when evidence of adaptation might appear weak, further inspection reveals that translators often restructure the message to a significant degree. Compared to other language pairs, translators from French into Russian seem to recognise that adaptation is required and make some effort to adjust the target texts accordingly to encode elements of the sightseeing phase.

7.3.4 Translations into Russian from Greek

The final language pair examined here is Greek into Russian, where a high degree of adaptation might be expected, as a result of the statistically significant differences between Greek and Russian non-translated texts regarding all semantic categories examined here, and a significant effort towards adaptation is noted. Successful adaptation is observed with Motion and Perception verbs, while weak adaptation is noted with Discovery and Emotion verbs. Figure 7.10 summarises the quantitative findings presented earlier in Table 7.8.

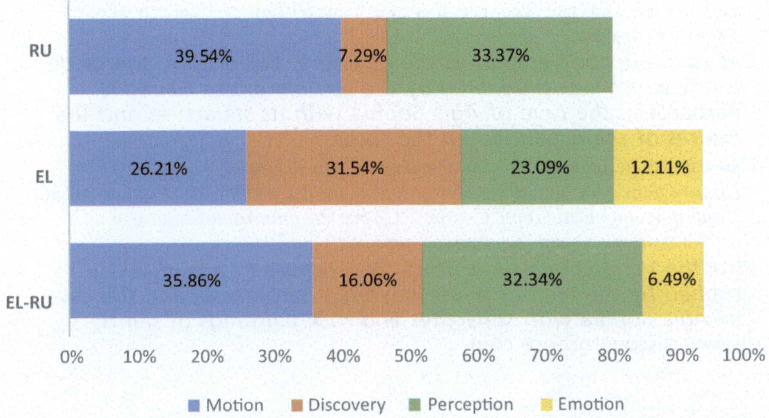

Fig. 7.10 Normalised frequencies of most prominent semantic categories non-translated Russian, Greek source texts, and translated Russian from Greek as % of all identified verbs

Motion verbs represent a higher frequency of verbs in target texts (35.86%) than in source texts (26.21%), close to what is found in Russian non-translated texts (39.54%). Similarly to what happens in French-Russian tourism translation, the significance of movement through space appears to be recognised, which is in line with how Russian non-translated tourism texts conceptualise tourist behaviour and tourism more generally. Example 7.22 illustrates how Motion verbs are added during translation.

Example 7.22

ST ***Γνωρίστε*** *το ενετικό κάστρο της Χώρας, την Παλαιοχώρα, την καστροπολιτεία, που λεηλατήθηκε από τον φοβερό πειρατή Χαϊρεντίν Μπαρμπαρόσα, το σπήλαιο της Αγίας Σοφίας με τους σταλακτίτες και τις μορφές των αγίων ζωγραφισμένες στους βράχους.*

BT ***Get to know*** *the Venetian castle of Chora, Paleochora, the castle town, which was plundered by the terrible pirate Hairedin Barbarossa, the cave of Agia Sophia with its stalactites and the figures of saints painted on the rocks.*

TT **Посетите** *венецианский замок Хоры "Палеохора" - город-крепость, ограбленный знаменитым пиратом Хайредином Барбароссой, и пещеру Святой Софьи со сталактитами и наскальными изображениями святых.*

BT ***Visit*** *the Venetian castle of Hora "Paleochora" - a walled city robbed by the famous pirate Hayredin Barbarossa, and the cave of Agia Sophia with stalactites and rock paintings of saints.*

Source www.discovergreece.com

In this example, the Discovery verb *γνωρίζω* (*get to know*) with a meaning close to *explore* is replaced with the Motion verb *посетить* (*visit*) during translation. There is a clear shift of emphasis from exploration to more prototypical elements of tourism, such as movement. In the Greek source text, the reader-tourist is invited to engage with the destination in an effort to connect with it, while in the Russian target text a more distanced notion of tourism is promoted, one where the only requirement is the reader-tourist's presence at the destination. However, this shift is successful as it allows the text to recreate the image of tourism that is compatible with the sightseeing phase of tourism that the Russian society is currently experiencing.

A similar pattern is observed in the case of Perception verbs, whose frequency is higher in target texts (32.34%), than in source texts (23.09%), resembling the patterns found in non-translated Russian texts (33.37%). The significance of Perception verbs, especially those relating to sight, appears to have been recognised during translation, in line with what is observed in Russian non-translated tourism texts. As before, verbs might be added during translation, as can be seen in Example 7.23.

Example 7.23

ST *Στη διαδρομή σας στο φαράγγι θα* ***συναντήσετε*** *ένα κατακόρυφο πέρασμα βάθους εννέα μέτρων όπου έχει τοποθετηθεί μεταλλική σκάλα για διευκόλυνση.*

BT *On your way to the gorge you will* ***encounter*** *a nine meter deep vertical passage where a metal ladder has been placed for convenience.*

TT *Когда вы будете идти вдоль ущелья, вы* **увидите** *9-метровый вертикальный проход с железной лестницей, которая была расположена там для удобства и в целях безопасности.*

BT *As you walk along the gorge, you* ***will see*** *a 9-meter vertical passage with an iron ladder that was placed there for convenience and safety.*

Source www.incrediblecrete.gr

Here, the Discovery verb *συναντώ* (*encounter*), which could have been translated as *находить* (*find*), is instead translated using a Perception verb, and more specifically one related to sight *увидеть* (*see*). Similar to the previous example, there is a clear shift here from exploration to more prototypical elements of tourism, such as sight. It is also interesting to note in this example that the opening noun phrase *στη διαδρομή σας* (*on your way*) is transformed into a verb phrase where the reader-tourist is the subject, i.e. *когда вы будете идти* (*as you walk*). This is another example of more emphasis being placed on Motion verbs. Even though the meaning conveyed might be the same (after all the Greek source text implies that the walking is done by the reader-tourist though the use of the possessive pronoun), the message is more explicit in the Russian target text.

Further evidence of successful adaptation can be obtained from examining the verbs used to conceptualise Perception and specifically the fact that in Russian translations no mention is found of experiencing a destination, although some references to that element of tourism have been noted in Greek tourism texts mainly through the use of the verb *ζω* (*live*). In the target texts, the verb *почувствовать* (*feel*) is occasionally found which is, however, not as strong as *испытывать* (*experience*) or *погружаться* (*immerse*). Some attempt at least is, therefore, made to negotiate the differences between Greek and Russian tourism discourse during translation. Example 7.24 illustrates a successful instance of adaptation in that regard: the experience-related Perception verb *ζω* (*live*) is replaced by a Possession verb *проводить* (*spend*) in the target text focusing on spending time rather than on sensations.

Example 7.24

ST *Τα νησιά του Ιονίου και του Αιγαίου και οι ακτογραμμές της ηπειρωτικής Ελλάδας σας περιμένουν, για να* ***ζήσετε*** *το ελληνικό καλοκαίρι και να χαρείτε τον ήλιο και τη θάλασσα μαζί με τους αγαπημένους σας.*

BT *The islands of the Ionian and Aegean and the coastlines of mainland Greece are waiting for you, to* ***experience*** *the Greek summer and enjoy the sun and the sea with your loved ones.*

TT *Острова Ионического и Эгейского морей, побережье материковой Греции ждут вас, чтобы вы* **провели** *здесь греческое лето и насладились солнцем и морем вместе со своими близкими людьми.*

BT *The islands of the Ionian and Aegean Seas, the coast of mainland Greece are waiting for you to* ***spend*** *the Greek summer here and enjoy the sun and the sea with your loved ones.*

Source www.discovergreece.com

It is not a coincidence that in Examples 7.22 and 7.23, the Greek source text employs a Discovery verb, since weak adaptation is observed in these cases. Discovery verbs are strongly preferred in Greek tourism texts (31.54%), but they are used less frequently in the Russian translations (16.06%), coming closer to frequencies found in Russian non-translated tourism texts (7.29%) but not matching them. Examples 7.22 and 7.23 illustrate how this reduction in the frequency of Discovery verbs might manifest itself, i.e. by a shift in the semantic category of the verb. However, Discovery verbs might also be adapted by omission, as can be seen in Example 7.25.

Example 7.25

ST *Ό,τι και αν* ***αναζητάτε,*** *η Κάλυμνος είναι ένας σπάνιος θησαυρός, ένα κομμάτι της αυθεντικής ψυχής της Ελλάδας που πρέπει οπωσδήποτε να* ***ανακαλύψετε.***

BT *Whatever* ***you are looking for,*** *Kalymnos is a rare treasure, a piece of the authentic soul of Greece that you must* ***discover.***

TT *Калимнос – это редкое сокровище, маленький кусочек подлинной греческой души, который вы обязательно должны* **открыть** *для себя!.*

BT *Kalymnos is a rare treasure, a little piece of the authentic Greek soul that you should definitely* ***discover!***

Source www.discovergreece.com

In this example, the first part of the clause where the Discovery verb *αναζητώ* (*look for*) is found is omitted in the target text. However, the other Discovery verb *ανακαλύπτω (discover)* at the end of the clause is translated as *открыть* (*discover*) in Russian. The different approaches to translating Discovery verbs illustrated in these examples, that is semantic shift, omission, and retention, suggest a mixed approach which contributes to the weak adaptation pattern. This results in more emphasis being placed on actively searching the destination in translated Russian texts from Greek compared to Russian non-translated tourism discourse, reflecting elements of the post-sightseeing tourism phase.

The adaptation of Discovery verbs might however be more successful than it seems. The examination of the verbs used to express this semantic category reveals that no emphasis is put on exploration through verbs such as *исследовать* (*explore*), even though their equivalents are found in the Greek source texts. Such verbs are often omitted as can be seen in Example 7.26. This suggests that translators have successfully recognised the need to focus more on the outcome of the exploration in Russian tourism discourse in accordance with the sightseeing phase of tourism that the Russian society is currently experiencing.

Example 7.26

ST ***Εξερευνήστε*** *το Αργοστόλι, τη γραφική πρωτεύουσα.*
BT ***Explore*** *Argostoli, the picturesque capital.*
TT *Аргостоли: столица и гавань.*
BT *Argostoli: the capital and harbor.*
Source www.discovergreece.com

Finally, weak adaptation is also observed with Emotion verbs. These verbs are found in Greek source texts (12.11%), and they are also found in Russian target texts but with a lower frequency (6.49%), even though a much higher degree of adaptation would have been expected to match what is found in Russian non-translated tourism texts where these verbs have not been identified. The main strategy through which this is achieved is omission, as seen in Example 7.27.

Example 7.27

ST *Εδώ θα βρεις πολλές ταράτσες για να* ***απολαύσεις*** *τα κοκτέιλ ή το ρομαντικό σου δείπνο με θέα την Ακρόπολη.*
BT *Here you will find many terraces to* ***enjoy*** *cocktails or a romantic dinner overlooking the Acropolis.*
TT *Здесь Вы сможете найти широкий выбор баров с террасами для вечернего коктейля или романтического ужина с видом на Акрополь.*
BT *Here you can find a wide range of bars with terraces for an evening cocktail or a romantic dinner overlooking the Acropolis.*
Source www.discovergreece.com

In Example 7.27, the Emotion verb *απολαμβάνω* (*enjoy*) has been omitted from the target text together with the reference to the reader-tourist. This is in accordance with the communicative conventions of the Russian tourism discourse and this particular instance can be considered a successfully adapted one. However, a more widespread use of this strategy would have been required to encode more clearly the sightseeing phase that the Russian society is currently experiencing. For both Discovery and Emotion verbs, the target text encodes some elements of the post-sightseeing phase although a clear attempt is noted, especially in the case of Discovery verbs, to tone such elements down as much as possible. While there is certainly scope for further adaptation here, it cannot be ignored that translators appear to have some understanding of which elements need to be adapted.

The examination of the final pair of examples provides evidence that translators from Greek into Russian have recognised the importance of adapting the tourism message to encode the concept of tourist behaviour and tourism more generally that is circulating in the Russian society and subsequently to make the tourism message more appealing to Russian-speaking tourist-readers. Even though adaptation is not always successful for all semantic categories, the resulting text as a whole is much closer to the Russian tourism discourse than the Greek one.

Overall, a stark contrast in adaptation patterns is observed between Russian texts translated from English, irrespective of the variety, and from the other two languages examined here. Adaptation is higher when the source texts are French or Greek compared to English, even though a similar degree of adaptation might be expected in all cases. Why such a high degree of adaptation is noted with these particular language pairs is not clear, but it might have to do with the status of these languages compared to English.

7.4 Discussion

The examination of adaptation patterns have revealed a rich canvas of different strategies and adaptation patterns, some of which might be more easily explained than others. Nevertheless, even when adaptation has been found to be successful, a careful examination of the concordance lines reveals that not all aspects of tourist behaviour have always been adapted. The reverse is also true: specific aspects of tourist behaviour, as evidenced by the use of specific verbs, might have been adapted, but not the overall frequency of the semantic category these belong to. Overall, results reveal that there are different approaches to translation, which can be roughly divided into three groups: no adaptation, that is, reluctance to move away from source text; adaptation, which might vary between weak and successful; and lack of a clear pattern, where adaptation appears to be inconsistent, especially where additional adaptation occurs even in cases where it might not be required.

In the first group, that of no adaptation, we mainly find texts translated from American English into French and Russian, as well as English translations of French tourism texts and Greek translations into French. However, not all cases should be considered equal, since adaptation is not expected to the same degree across all these language pairs. For example, there are many more differences between American English and Russian tourism texts, and consequently a much higher degree of adaptation is expected, compared to Greek translations into French.

The reasons for the reluctance to move away from the source text might be multiple, but data suggest that a strong motivating factor is the ubiquitous role of Anglophone, mainly American, culture. Prestige therefore seems to be a hindering factor to adaptation and a motivating factor for source texts patterns to be adopted in translated texts. It is also possible that when low degrees of adaptation are required, i.e. when two societies experience more or less the same tourism phase, translators are more likely to perceive the tourism discourses in the two languages as similar and stay close to the source text. Other reasons for staying close to the source texts might involve instructions received by commissioners of these translations. Translators might be instructed to stay as close as possible to the source text and not make any changes of any kind. Whatever the reasons might be, and it is likely to be a complex interplay of factors, it reflects a general lack of understanding of the importance of adaptation of the tourism message, which is not limited to a specific locale or language pair.

In the second group, that of adaptation (at least to some extent), we find language pairs where Russian is either the source or the target language (except for when American tourism texts are translated into Russian, where generally no adaptation has been observed). Particular evidence of the generally high degrees of adaptation noted with Russian comes from the semantic category of Emotion. This category has not been found to be as frequently used in Russian tourism texts, and adaptation is most noticeable in this case. It is also only in translations into Russian that some attempt at adapting the verbs through which semantic categories are realised, is also observed.

What seems to encourage attempts towards adaptation is the degree of adaptation required. In all cases, a generally high degree is needed as a result of the differences between non-translated texts in Russian and the other languages examined here. When high degrees of adaptation are required, it is perhaps easier for translators to recognise the need for it and make an attempt to move towards it. This might be especially the case where the translation pair does not involve English and therefore the status of the languages does not affect the translation process. However, it cannot be argued with certainty to what extent adaptation in this case is a conscious attempt or the result of a subconscious internalised understanding of tourism discourse.

Finally, in the third group, that of no clear pattern, we find English translations of Greek texts and French translations of British English texts. These are the only language pairs where additional or over-adaptation have been noted. A lack of clear pattern once again demonstrates a general lack of understanding of the importance of adaptation, which is not restricted to a specific language pair. Translators often appear divided regarding the approach they should follow producing a target text that shows elements of different adaptation patterns. This also reflects the fact that translators might be lacking the knowledge of what adaptation should look like and what changes need to be introduced. This lack of consistency is a manifestation of the recognition that changes are required but also shows a lack of understanding of how they need to be implemented. This is why significantly more studies like the present one are needed to shed light on the subtle and not so subtle cross-linguistic differences of tourism discourse.

Although there are some successful attempts at adaptation, what the data largely point towards is a generalised lack of understanding of the importance of adapting the tourism message when addressing a new audience and where indicated by the different tourism phase experienced. It is of course no surprise that translation is more than the transfer of words, but by examining tourism translation, we can gather evidence that translation can represent a certain version of the world that might or might not correspond to that of the target readers. While it is easy to put the blame on translators for this state of affairs, it is actually the case that we know very little about the power of tourism discourse, how it can reflect the way we view the world, and how this can be used to inform translation practice. The present study has aimed to address this shortcoming by providing a clear rationale in the form of the tourism phase framework for why cross-linguistic differences might be observed. Linguistic differences are therefore not simple idiosyncrasies of tourism texts in different languages but can be explained and are linked to each other, which hopefully makes it much easier to introduce changes during translation in the right direction.

Furthermore, results from the study on tourist behaviour suggest that translation has a significant potential to increase or decrease the promotional potential of the tourism message. A message that is in line with how reader-tourists perceive tourism is likely to create a more positive image of the destination in their minds, compared to one that contrasts with it. Staying close to the source text, for instance, or creating an idiosyncratic tourism text that belongs to neither the source nor the target culture is likely to compromise the promotional effect of the tourism message and not turn readers into visitors. That is why it is important to have an in-depth understanding of the target society and what it prioritises in the tourism experience. This can be achieved in many ways, from marketing research and questionnaires (by destination promoters) to a close examination of the communicative conventions of non-translated tourism texts in the target language (by translators).

Tourism is not simply about attracting visitors to a place. It is a unique window on how we view the world around us. Our perceptions of the world are encoded in language and what better text to do that than one that describes the world around us and how we might interact with it. Whether a culture puts emphasis on emotions in tourism or whether it focuses mainly on movement can tell us a lot about how particular societies might operate and what their values are. Whether the focus is on active exploration or not, can reveal the role that citizens play in their society. The present study is not a cultural exploration of these issues, however. Instead, the focus is on how translation can bridge the gap between alternative worldviews. And there is evidence from the examination of tourist behaviour that translation certainly has that potential by encoding a different phase of tourism, for example. To further explore this finding, the second important component of tourism will be examined next, that is the places to go, which forms part of the second case study of this book.

Part II

Case study II: Places to Go

8

Case Study II: Key Concepts and Approach

The second case study focuses on the other key element of tourism discourse, that is, places to go. More specifically, it takes the destination, and its elements, as a point of departure: the focus is on *places*, or rather physical space, which is understood here is in a broad sense to refer to either natural (e.g. island, mountain) or urban (e.g. city, park) landscape. Tourism texts promote certain aspects of physical space through repetition but importantly by describing it in, often overtly, positive terms. This aspect of tourism texts is perhaps what brings them the closest to advertising discourse, where there is also a tendency to use positive language for persuasion. What is more, evaluation, which is typically associated with the use of adjectives, is frequently identified as a distinguishing feature of the language of tourism (see Chapter 2). For those reasons, the focus in this exploration is on adjectives that modify nouns referring to physical space.

The main assumption is that readers with different linguistic and cultural backgrounds are likely to be used to perceiving physical space differently, and therefore accustomed to different descriptions of it when it comes to its promotion. This process plays a key role in destination image formation, as it allows for distinct imageries to be created for

S. Malamatidou, *Translating Tourism*, Palgrave Studies in Translating and Interpreting, https://doi.org/10.1007/978-3-031-49349-2_8

different linguistic groups. Therefore, this case study will reveal how places are described and promoted in tourism texts and their translations, and also how different linguistic groups perceive and classify a destination, which will provide important insights into how we view physical space, and how this might differ across languages.

To establish possible differences, non-translated tourism texts in American and British English, French, Greek, and Russian are examined first. It is assumed that patterns found in non-translated texts reflect the way different societies perceive physical space and how this is encoded into linguistic structures and word choices in each language. In addition, these patterns represent the communicative preferences to which readers in different languages are typically exposed. When it comes to translation, particularly between languages that exhibit different communicative preferences, it is assumed that successful promotion is achieved when translated texts are adapted where necessary to the preferences of the target language to which target readers are accustomed, and which reflect their way of gazing at the world around them. Therefore, the main aim of this exploration is to establish whether and how successfully this adaptation occurs.

8.1 Physical Space

To understand why and how people might perceive the world around them differently, it is important to first consider the importance of physical space in tourism texts, and tourism more generally. Tourism cannot exist without physical space, and tourism texts are essentially about the promotion of a specific geographical area and key elements of it. For destination promotion to be successful, it is key that physical space is marketed as unique, and therefore it is important that tourism texts and other promotional material focus on presenting aspects of physical space in a way that will make it stand out. This process is complex and results in geographical locations which are often very similar to one another becoming tourism destinations by being presented as unique (Hughes 1998). Therefore, destination promotion is achieved by semiologically differentiating a place; a carefully designed and self-reflective process (ibid.) in which destination promoters consciously engage. As a result, a curious paradox is observed: experience suggests that tourism destinations have become in many ways similar under the influence of market forces and globalisation, while the language used to promote them aims at distinctiveness. Another related paradox is that of destinations being often described as 'hidden gems' or 'less crowded' in order to attract even more visitors, therefore ultimately losing what made them attractive in the first place. Tourism discourse has the power to transform the image of a physical space both literally and metaphorically.

Space can be understood in many different ways, for example from a geographical, sociological, or tourism perspective, and therefore a single definition is hard to find. For instance, a tourism definition of space postulates that a tourist is a prerequisite (Włodarczyk 2009), while a geographical definition of space promotes an idea of space as constantly changing as a result of the elements that interact within it, rather than as a pre-existing space where things are simply embedded (Thrift 2009). The more one tries to understand and define space, the more one realises how complex, yet fundamental, a concept it is, to the extent that Hall, a well-known anthropologist, once compared it to sex by arguing that "[i]t is there but we don't talk about it. And if we do, we certainly are not expected to get technical or serious with it" (cited in Barcan and Buchanan 1999, 7). The aim here is not to offer an in-depth account of what space is but rather to come up with an understanding of physical space that will help guide this case study. Therefore, physical space is understood here as any element of the landscape of a geographical area, including the area itself. In other words, any aspect of space that is presented in tourism texts as worth visiting, seeing, or experiencing. This might refer to the natural landscape, such as an island, beach, mountain, river, or an urban landscape, such as a city, town, park, street, bridge. Buildings and specific attractions, e.g. museums and galleries, are also considered elements of space, as are restaurants and bars. Instances of more 'abstract' references to space, such as area or place, are also included. From a methodological point of view, this rather broad understanding of space increases the number of adjectives captured.

8.2 Tourist Gaze

How does an apparently ordinary place get transformed into a tourist destination through the power of language? How do destination promoters decide which elements of it to focus on and promote? As already mentioned in Chapter 2, according to research conducted by Urry and Larsen (2011), central to tourism is the idea of gaze, as 'site' needs to be transformed into 'sight'. Tourist gaze suggests that tourism experience involves a distinct way of seeing; it is an idea associated with the construction of the tourism experience and all the dynamics and processes associated with it. It refers to a socially informed and technologically mediated gaze, directing tourists to gaze upon places that are out of the ordinary (Urry and Larsen 2011). Tourism is a way of seeing, with destination promoters offering visual experiences and visitors consuming these. Of course, this focus on the visual reflects the general predominance of sight particularly in Western societies (Hughes 2012), without ignoring the role that other senses might play. Although tourism is a multisensory activity, sight plays a key role, manifested not least from the exponential growth of photos taken and shared on social media. Based on this idea of the tourist gaze, tourism is about organising and transforming places as sights and gazing is about consuming these visually (Urry 1995).

Unlike eyesight, the tourist gaze is not something that most humans are born with, but rather the result of a learned way of seeing, and many sources can inform tourist gaze over time, from authors to social media, and from hotel owners to TV travel programmes. Urry and Larsen make it very clear that "[t]here is no single tourist gaze as such" and that "the pure and innocent eye is a myth" (2011, 1–2). According to Jenks "[t]here is nothing 'out-there' intrinsically formed, interesting, good or beautiful", instead "[v]ision is skilled cultural practice" (1995, 10). Our eyes, Urry and Larsen (2011) argue, are socio-culturally framed and there are many different ways of seeing. Tourist gaze is the way of seeing and interpreting new places, and is framed by social class, gender, nationality, age, and education, which are factors that also affect promotions as discussed in Chapter 2. There are many ways of gazing and, as a result, our perception of a destination is subjective (Mayo and Jarvis

1981). This gaze is naturally encoded and reflected in the discourse of tourism, which is where the destination is "unfailingly inscribed in the glance which discursively *defines* [it]" (Baider et al. 2005, 27, emphasis in the original).

All this means that we are consumers of culturally constructed imageries, whose aim is not to reflect the world, but rather to order, shape, and organise it in our minds. This process feeds into and is fed by socio-cultural preferences, giving tourism texts not only a strong promotional function but also an ideological one. Every time a new place is presented, we approach it with pre-established notions about physical space conditioned by our personal experiences and memories and framed by rules and styles, as well as previous imageries and texts. These frames can transform physical space into something that is interesting, good, or beautiful and, without them, the world would appear to be a very different place. As with other aspects of tourism, this creates a curious paradox: tourists tend to gaze upon places that they have already consumed in some way either in image or text form, sometimes unbeknownst to them.

It is possible to understand how tourist gaze is constructed in different linguistic societies and how tourism texts transform site into sight by focusing on how physical space is described. Translators are expected to (re)create imageries of physical space that are, ideally, in accordance with the readers' viewpoints and are successful in directing their gaze towards those aspects of the physical space that they are conditioned to gaze at. Translation, in this case, is about transferring attitudes and perspectives, and translators need to be able to look at the world through someone else's eyes. This concept of tourist gaze is distinct from tourism phase, although the two can be related. Tourist gaze focuses more directly on visual experiences of destinations and is therefore more appropriate as a concept for examining physical space, while tourism phase captures a wider number of things related to the tourism activity. A specific tourist gaze can be associated with a certain phase of tourism, as will be shown in Chapters 9 and 10, but more research is required to understand how exactly these two concepts interact and overlap with each other.

8.3 Methods of Analysis

With a better understanding of physical space and tourist gaze, it is now time to turn our attention to how it can be encoded linguistically. Although the framing of physical space can be linguistically achieved in different ways, the most obvious one is through descriptions, typically associated with the use of adjectives. Adjectives play a key role in many registers and even more so in tourism texts. They are used to modify nouns and can express the writer's attitude and shape the reader's response to the text, while they are also responsible for the communication of key ideas in the text.

Adjectives are chosen here instead of descriptions more generally because they can provide a rich category for analysis, since they play a key role in tourism discourse, and promotional language more generally, while they offer methodological advantages when it comes to their identification using corpus methods, that is, they can be easily captured. Studies on tourism texts also link adjectives to the language of euphoria. For instance, Gotti (2006) argues that the language of tourism is typically evaluative, and that there is a recognised tendency of tourism texts to speak only in positive terms of the destination they seek to promote. As a result, particular emphasis is placed in the current empirical investigation on how evaluation is achieved linguistically through the use of adjectives.

However, even with this focus in mind, the number of adjectives identified by SketchEngine within each corpus component is very large, making it hard to study all adjectives in a corpus of this size. Instead, as explained in Chapter 4, the focus here is on the most frequent adjectives in each corpus component meeting the stated criteria and after the cut-off point of 0.18 ptw has been applied (for reasons explained in Chapter 4). It is worth recalling here how these 'top' adjectives have been identified. Firstly, a lemmatised wordlist of adjectives sorted by frequency was generated for each corpus component with the help of SketchEngine (Kilgarriff et al. 2014), which allows for uploaded corpora to be automatically POS tagged, significantly facilitating the identification of adjectives. Following this, the concordance list for each adjective (generated by selecting each adjective from the lemmatised frequency list) was examined separately to understand the context of each clause and establish whether what is being modified by the adjective in question corresponds to physical space. Both pre-modification and post-modification were considered, the latter being the norm in French. Similarly, both attribution (e.g. *the blue sea*) and predication (e.g. *the sea is blue*) were examined, although the latter was found to be rare. Concordance lines where physical space is not modified by the adjective were excluded from the analysis.

During the examination of concordance lines, each adjective was analysed to identify potential instances of polysemy (e.g. *top* can mean both the best and the highest part of something), since the focus is on meaning rather than form. Such instances were generally rare. This resulted in a list of adjectives and their raw and normalised frequencies based on their meaning for each corpus component. The next step was to assign the adjectives to semantic categories to support the analysis.

The classification of adjectives is largely based on Biber et al.'s (1999) model. Using a corpus-based approach to English grammar, where the grammatical descriptions are based on linguistic patterns actually used by speakers and writers and found in a large (i.e. over 40-million-word) electronic collection of texts, they focus on the semantic properties of adjectives and divide them into two broad groups: descriptors and classifiers. In the analysis that follows in the next two chapters, the focus will be on descriptors only for reasons that will be explained in detail in Chapter 9.

Descriptors denote characteristics such as colour, size and weight, chronology, emotion, and a wide range of other features which belong to the miscellaneous category. The distinguishing characteristic of descriptors is that they are gradable, although it must be noted that some descriptors can be more easily graded than others (compare, for instance, *very big* and *very enormous*; whereas both are descriptors denoting size, the former is more often graded than the latter). **Classifiers** describe an entity in relation to other referents and they are distinguished from descriptors because they are typically non-gradable. These two broad semantic categories can be further divided into subgroups as follows:

Descriptors

- **Colour**: These are adjectives denoting colour or brightness, such as *blue, white, bright, dark.*
- **Size/Quantity/Extent**: These are adjectives denoting size, weight, or extent, such as *big, tiny, heavy, light, long, narrow.*
- **Time**: These are adjectives denoting chronology, age, and frequency, such as *modern, recent, new, old, late.*
- **Evaluative/Emotive**: These are adjectives denoting judgement, affect, and emphasis, such as *bad, lovely, fantastic, poor, beautiful.*
- **Miscellaneous descriptive**: This is a very broad subgroup which includes all other gradable adjectives that do not belong to any of the above subgroups, such as *exotic, strong, popular, different.*

Classifiers

- **Relational/Classificational/Restrictive**: These are adjectives that delimit the referent to a noun, particularly in relation to other referents, such as *main, previous, standard, public, south.*
- **Affiliative**: These are adjectives denoting a national or religious group, or any other group, such as *British, Catholic, United.*
- **Topical/Other**: These are adjectives denoting the subject area or showing a relationship with a noun or any other classifier that does not belong to any of the above subgroups, such as *visual, volcanic, medieval, coastal.*

This framework offers a certain level of detail and aims at some degree of objectivity, which significantly facilitates categorisation and analysis. However, there is still some degree of subjectivity which means that each concordance line needs to be examined carefully. For instance, Biber et al. (1999) admit that some adjectives can serve as either descriptors or classifiers depending on their meaning and the context in which they are used (i.e. the noun that they modify), such as *modern,* which can denote something that is currently used, as opposed to an earlier form (e.g. *modern English*) in which case it is a classifier, or it can denote something relating to present or recent times (e.g. *modern technique*) in which case it is a descriptor. The same can be said for adjectives such as *old*, which can denote either age or evaluation, and *poor*, which can be either an emotive or a miscellaneous descriptor. However, Biber et al. recognise that the different meanings are often associated with register, and therefore in a study of tourism texts, it is expected that the same adjective will, in the majority of cases, be used with the same meaning, especially if it modifies nouns belonging to the same semantic 'set', that is, physical space.

What is more, Biber et al.'s (1999) semantic categorisation of adjectives allows for clearer links between a certain semantic category of adjectives and the promotional effect of tourism texts. Descriptors are more likely to be used to create a positive image of a destination, directing the reader's gaze and attention to those aspects of physical space that make it stand out. For instance, describing a destination using a descriptor such as *beautiful* or *popular* has a more direct effect on the creation of "a positive image of a destination as a holiday attraction worth visiting, beautifying and celebrating physical space" (Francesconi 2014, 58), compared to classifiers like *cultural* or *nearby*. This is not to say that classifiers do not contribute towards the promotional function of tourism texts, but rather that their potential in that regard is limited compared to descriptors. The gradability of descriptors is also what allows for superlatives, another typical feature of tourism texts, which contributes towards their promotional effect. For this reason, more emphasis is placed in this empirical investigation on descriptors compared to classifiers. Also, as will be seen in the following chapters, descriptors are more frequently found in tourism texts than classifiers, further supporting the idea that they are more prominent in this discourse.

Despite the significant descriptive advantages of this framework, there are some aspects of it that remain vague, which make its application to data challenging. One such aspect is gradability, which can in some, admittedly limited cases, be applied to classifiers as well (e.g. *a very British dish*, where *British* is an affiliative classifier which is normally non-gradable), while, as has already been mentioned, some descriptors are 'more gradable' than others. Additionally, it is not always clear how Biber et al. (1999) categorise some adjectives if the gradability principle is to be followed. For instance, they mention the adjective *different* as an example of a classifier, and it is indeed possible to argue that this adjective requires another referent and an element of comparison (i.e. different from something), and is therefore relational. However, this adjective demonstrates a very strong preference towards gradability and therefore would be a good candidate for a descriptor, which is how it is treated in the analysis of the data, following the gradability principle. Similarly, *former*, which at first sight might appear to be a classifier (also following the gradability principle), is considered to be a descriptor here as it is typically used with the sense of old.

Last but not least, the two subgroups of evaluative/emotive adjectives and miscellaneous ones are rather loosely defined. On the one hand, there is not much explanation as to what judgement, affect, or emphasis, which is what evaluative/emotive adjectives are said to denote, refer to exactly and why they have all been grouped together. On the other hand, the miscellaneous category is just too broad to be practicable and lacks the specificity of the remaining categories. Despite these limitations, Biber et al.'s framework can still be useful for the classification of adjectives and will be adopted in the present case study with a few adaptations, namely the inclusion of more semantic categories based on the data from the TrAIL corpus, and the use of the gradability principle in conjunction with the sense of the adjective under examination (see, for example, discussion of *different* and *former* earlier). It is worth noting that what is included in each semantic category largely depends on how adjectives belonging to it are used in tourism discourse, and not how they might be used in everyday language or other types of discourse.

Place-names used as pre-modifiers, such as *Oxford city* or *Ionian islands* were excluded from the analysis as their main function is to name rather than describe. Similarly, all numerals (e.g. *100*) were excluded, as well as any linguistic items related to quantity (e.g. *many, numerous, first*) as they tend to act as determiners or pronouns. Even when these are used as adjectives, they seem to constitute a separate category, which incidentally Halliday and Matthiessen (2014) call numeratives and are not accounted for in Biber et al.'s (1999) semantic group model of adjectives.

Once all adjectives are grouped into semantic categories, it is possible to identify which aspects of the destination are stressed in each language/

type of text (non-translated/translated) by examining the semantic categories that are more (or less) prominent. This can in turn provide information about how different societies gaze at the world around them. Table 8.1 presents the semantic categories identified in the TrAIL corpus and the adjectives through which they have been found to be realised across all languages and types of texts. The analysis then focuses on the overall distribution of the categories and how they are realised through specific adjectives. Any comparisons made across corpus components are based on statistical analyses using the log-likelihood text and Bayes (BIC) factor (see also Chapter 4 for details on methodology). As explained in Chapter 5 relating to the first case study, whenever a statistical difference is reported when presenting the results in the next two chapters, this means that the probability value p has been found to be less than 0.05 and that the BIC factor has been found to be larger than 2. Also, as in Chapters 6 and 7, the detailed results of statistical tests will not be reported, especially considering how many components need to be compared. In total, 33 comparisons have been made across five semantic categories which were found to be most prominent (see Chapter 9): ten comparisons involving non-translated texts, nine comparisons involving texts translated into English, eight involving texts translated into French, and six involving texts translated into Russian. If each semantic category is counted as a separate comparison, there were altogether 165 comparisons. The full details of all statistical tests can be found in Appendix II.

As with the previous case study, semantic categories are not measured based on normalised frequencies for each category based on all words in the corpus component, but instead based on the number of identified adjectives above the cut-off point (e.g. Time adjectives represent 10% of all identified adjectives). This is because normalised frequencies are likely to simply reflect the differences in the total number of adjectives identified, which might be quite significant across languages. For example, a low normalised frequency of Quality adjectives in American English tourism texts might simply reflect the fact that fewer adjectives (compared to texts in the other languages) modifying nouns denoting physical space are used in those texts, overlooking that in fact Quality adjectives are one of the most prominent categories in American English

Table 8.1 Most frequent adjectives modifying nouns denoting physical space in all corpus components grouped into semantic categories by language

Group	Semantic category	English	French	Greek	Russian
Descriptors	Colour	green, white, golden	vert (green), turquoise (turquoise)	καταπράσινος (green)	зеленый (green), белый (white), красный (red), бирюзовой (turquoise)
	Size(also includes extent)	large, small, high, big, little, vast, narrow	grand (large), petit (small), haut (high), vaste (vast), escarpé (steep), étroit (narrow), profond (deep)	μεγάλος (large), μικρός (small), ψηλός (high), στενός (narrow), τεράστιος (huge)	большой (large), крупный (big), небольшой (small), просторный (spacious), высокий (high), крупный (large), обширных (extensive), бескрайний (boundless), огромный (huge), извилистый (winding), узкий (narrow), маленький (small), глубокий (deep)
	Time	old, ancient, former, traditional, new, modern	ancien (ancient), vieil (old), antique (ancient), traditionnel (traditional)	αρχαίος (ancient), παλιός (old), παραδοσιακός (traditional), σύγχρονος (modern)	новый (new), старый (old), бывший (former), современный (modern), старинный (vintage), древний (ancient), модный (trendy), традиционный (traditional)

Group	Semantic category	English	French	Greek	Russian
	Beauty	beautiful, scenic, stunning, picturesque, magnificent, spectacular, charming, pretty, wonderful	beau (beautiful), magnifique (magnificent), pittoresque (picturesque), charmant (charming), joli (pretty)	γραφικός (picturesque), όμορφος (beautiful), υπέροχος (wonderful), πανέμορφος (gorgeous), ωραίος (beautiful)	живописный (picturesque), красивый (beautiful), прекрасный (beautiful), очаровательный (charming), великолепный (magnificent), колоритный (picturesque), чудесный (wonderful), сказочный (fabulous), райский (heavenly)
	Quality	good, perfect, top, great, ideal, exceptional	bon (good), parfait (perfect), superbe (superb), idéale (ideal)	ιδανικός (ideal), καλός (good)	хороший (good), идеальный (perfect), отличный (great)
	Effect	dramatic, impressive	imposant (imposing), impressionnant (impressive), spectaculaire (spectacular)	εντυπωσιακός (impressive), επιβλητικός (imposing), μαγευτικός (enchanting)	величественный (majestic), удивительный (amazing), впечатляющий (impressive)
	Distinctiveness	unique, quirky, cosmopolitan	particulier (special), insolite (unusual), unique (unique), cosmopolite (cosmopolitan), exotique (exotic)	μοναδικός (unique), κοσμοπολίτικος (cosmopolitan), εξωτικός (exotic)	уникальный (unique), единственный (unique), экзотический (exotic)

(continued)

Table 8.1 (continued)

Group	Semantic category	English	French	Greek	Russian
	Fame	popular, famous, iconic, legendary	célèbre (famous), mythique (mythical), emblématique (iconic), populaire (popular)	σημαντικός (important), διάσημος (famous), δημοφιλής (popular), γνωστός (well-known), περίφημος (famous), σπουδαίος (important)	известный (famous), популярный (popular), знаменитый (celebrated), легендарный (legendary), важный (important)
	Other	lush, independent, cosy, tranquil, wild, authentic, clear, fascinating, elegant	sauvage (wild), incontournable (unmissable), luxuriant (lush), élégant (elegant), fascinant (fascinating), paisible (peaceful), verdoyant (lush), accessible (accessible), véritable (genuine)	πεντακάθαρος (spotless), κατάφυτος (verdant), πολυτελής (luxurious), ιερός (sacred)	роскошный (luxurious), пышный (lush), спокойны (calm), дикий (wild), уютный (cozy), шикарный (chic), оживленный (lively), необычный (unusual), настоящий (real), чистый (clean), организованный (organised), богатый (rich), интересный (interesting), любимый (favourite), священный (sacred), доступный (accessible), искусственный (artificial),

Group	Semantic category	English	French	Greek	Russian
Classifiers	Relational	nearby, other, main, central, other, west, main, east, northern, southern, eastern, western	autre (other), principal (main), sud (south)	κοντινός (nearby), νότιος (south), κεντρικός (central), ολόκληρος (entire), ομώνυμος (namesake), βόρειος (north), ανατολικός (eastern)	главный (main), северный (northern), разный (varied), основной (main), южный (southern), западный (western), главный (main), восточный (eastern), центральный (central), соседний (neighbouring), общественный (public)
	Affiliative	national, Scottish, French, Greek, Russian, Byzantine, Venetian	national (national), parisien (Parisian), français (French), grecque (Greek), vénitien (Venetian), byzantin (Byzantine)	ελληνικός (Greek), βυζαντινός (Byzantine), ενετικός (Venetian)	московский (Moscow), национальный (national), шотландский (Scottich), британский (British), английский (English), французский (French), романский (Romanesque), римский (Roman), греческий (Greek), византийский (Byzantine), венецианский (Venetian)

(continued)

Table 8.1 (continued)

Group	Semantic category	English	French	Greek	Russian
	Topical/Other	historic, local, natural, urban, coastal, cultural, free, medieval, rocky, archaeological, sandy, mountainous, neoclassical, wooden, architectural, outdoor	naturel (natural), historique (historical), touristique (touristic), médiéval (medieval), culturel (cultural), skiable (skiable), commercial (commerical), urbain (urban), archéologique (archaeological), thermale (thermal)	αρχαιολογικός (archaeological), ορεινός (mountainous), ιστορικός (historical), φυσικός (natural), πέτρινος (stone), μεσαιωνικός (medieval), νεοκλασικός (neoclassical), παραλιακός (coastal), παραθαλάσσιος (coastal), αμμώδης (sandy), βραχώδης (rocky), τουριστικός (touristic), κρυστάλλινος	деревянный (wooden), исторический (historic), государственный (state), торговый (commercial), природный (natural), каменный (stone), городской (city), смотровой (observational), горный (mountainous), мемориальный (memorial), выставочный (exhibition), песчаный (sandy), коралловый (coral), местный (local), художественный (artistic), тропический (tropical), прибрежный (coastal), ботанический (botanical), пешеходный (pedestrian), велосипедный (bicycle), средневековый (medieval), лесной (forest), сельский (rural), скалистый

Group	Semantic category	English	French	Greek	Russian
				(crystalline), ιαματικός (thermal)	(rocky), морской (maritime), музыкальный (musical), королевский (royal), туристический (touristic), портовый (port), загородный (country), береговой (coastal), холмистый (hilly), археологический (archaeological), морской (maritime), местный (local), горнолыжный, (ski), монастырский (monastic), туристический (touristic), высокогорный (alpine), рыбацкий (fishing), фахверковый (half- timbered)

tourism discourse when compared to the total number of identified adjectives. The focus here, however, is not on the number of occurrences, but on how semantic categories relate to each other to describe physical space, that is, on how the semantic categories are distributed in different components of the corpus.

In the following chapters, the analysis is divided into two distinct stages, each corresponding to a chapter. The first stage of analysis involves the examination of non-translated texts in American English, British English, French, Greek, and Russian to establish how physical space is described in these languages, and by extension how these societies order, shape, and organise the world around them. The second stage of analysis involves the examination of translated texts in English, French, and Russian out of several source languages depending on the availability of data in the TrAIL corpus to examine how any observed differences are negotiated during translation and whether target texts promote an image of tourism places in accordance with the target or source language. It is assumed that successful communication and promotion of the tourism message is achieved when translated texts are adapted to how the target society prioritises elements of the physical space and how it gazes at the world more generally. Adaptation is expected between societies that have been found to prioritise different elements of the physical space as encoded in different linguistic preferences. The more distinct societies are in their preferences, the higher the degree of adaptation expected (see Chapter 4 for more details on the types of adaptation examined).

References

Baider, F., M. Burger, and D. Goutsos. 2005. "Introduction." In *Tourist Communication: Discursive Approached to Identity and Otherness*, edited by F. Baider, M. Burger, and D. Goutsos, 27–43. Paris: L'Harmattan.

Barcan, R., and I. Buchanan. 1999. "Introduction: Imagining Space." In *Imagining Australian Space: Cultural Studies and Spatial Inquiry*, edited by R. Barcan and I. Buchanan, 1–7. Nedlands, Western Australia: University of Western Australia Press.

Biber, Douglas, Stig Johansson, Geoffrey Leech, and Susan Conrad. 1999. *Longman Grammar of Spoken and Written English*. Harlow: Longman.

Francesconi, Sabrina. 2014. *Reading Tourism Texts: A Multimodal Analysis*. Briston, Tonawanda, and Ontario: Channel View Publications.

Gotti, Maurizio. 2006. "The Language of Tourism as Specialized Discourse." In *Translating Tourism: Linguistic/Cultural Representations*, edited by Oriana Palusci and Sabrina Francesconi. Trento: Trento University Press.

Halliday, M. A. K., and M. I. M. Matthiessen. 2014. *Halliday's Introduction to Functional Grammar*. London & New York: Routledge.

Hughes, George. 1998. "Tourism and the Semiological Realization of Space." In *Destinations: Cultural Landscapes of Tourism*, edited by Greg Ringer, 17–32. London & New York: Routledge.

Hughes, Jason. 2012. "Editor's Introduction." In *SAGE Visual Methods*, edited by Jason Hughes, xxi–xiv. London: Sage.

Jenks, C. 1995. "'The Centrality of the Eye in Western Culture: An Introduction.'" In *Visual Culture*, edited by C. Jenks, 1–25. London & New York: Routledge.

Kilgarriff, Adam, Vít Baisa, Jan Bušta, Miloš Jakubíček, Vojtěch Kovář, Jan Michelfeit, Pavel Rychlý, and Vít Suchomel. 2014. "The Sketch Engine: Ten Years On." *Lexicography* 1: 7–36.

Mayo, Edward, and Lance Jarvis. 1981. *The Psychology of Leisure Travel: Effective Marketing and Selling Travel Services*. Boston: CBI.

Thrift, N. 2009. "Space: The Fundamental Stuff in Geography." In *Key Concepts in Geography*, edited by N. Clifford, S. Holloway, S. Rice, and G. Valentine, 85–96. London: Sage.

Urry, John. 1995. *Consuming Places*. 4th edition. London & New York: Routledge.

Urry, John, and Jonas Larsen. 2011. *The Tourist Gaze 3.0*. London: Sage.

Włodarczyk, B. 2009. "The Landscapes of Tourism Space." *Turyzm/Tourism* 19 (1): 83–90.

9

Physical Space

This chapter examines how physical space is described in non-translated texts to identify any cross-linguistic differences in the use of adjectives used to modify nouns referring to physical space. The focus is on specific semantic categories of adjectives which are most prominent in tourism discourse, their overall distribution, and how they are realised through specific linguistic choices, i.e. specific adjectives. This examination will reveal how each society, namely American, British, French, Greek, and Russian, gazes at the world around it, how it organises physical space, and which aspects of it are prioritised and promoted at least in the context of tourism discourse. Results from this stage of analysis will inform the analysis of translated texts in the next chapter.

9.1 Distribution of Adjectives

This step of the analysis serves as the necessary backdrop for the rest of the analysis in this chapter, that is, the total number of adjectives which meet the threshold frequency measure of 0.18ptw and which also modify a noun denoting physical space, will be used as the basis for calculating

S. Malamatidou, *Translating Tourism*, Palgrave Studies in Translating and Interpreting, https://doi.org/10.1007/978-3-031-49349-2_9

the frequency distribution of these adjectives across languages belonging to different semantic categories. Any differences in this overall distribution of the total number of identified adjectives, albeit interesting, will not be discussed in detail here, so as not to distract from the aim of the current case study, which is the examination of the tourist gaze as realised linguistically through descriptions of physical space.

Table 9.1 presents the frequency distribution of the most frequent adjectives (i.e. those that meet the threshold frequency measure of 0.18ptw and which also modify a noun denoting physical space). For consistency in the calculation of the threshold frequency, normalised frequencies (*nf*) are also calculated per thousand words. It is worth noting that the table captures adjectives across all the semantic categories identified in this case study; distinguishing between these will be applied at the next step of the analysis (Table 9.2). 'Tokens' refers to the total number of words in each corpus component (i.e. the collection of non-translated tourism texts in each language, see Table 4.1). 'Adjectives (types)' refers to the number of unique adjectives identified, irrespective of how many times they might be used in total, and 'Adjectives (tokens)' refers to the number of particular adjectives identified.

As can be seen from Table 9.1, there is indeed considerable variation across the languages examined as part of the TrAIL corpus regarding how frequently they describe physical space. Greek tourism texts use more adjectives to describe physical space compared to the tourism texts in the other languages examined here, while American English tourism texts employ these rather infrequently. Tourism texts in British English sit somewhere in-between, while French and Russian tourism texts are

Table 9.1 Normalised frequencies of most frequent adjectives modifying nouns denoting physical space in non-translated corpus components by language

	Tokens	*Adjectives (types)*	*Adjectives (tokens)*	
			f	*nf (ptw)*
EN_US	170,520	27	1,587	9.30
EN_UK	170,704	47	2,637	15.44
FR	170,060	29	1,791	10.53
EL	170,601	57	3,812	22.34
RU	170,881	34	1,892	11.07

closer to American ones. The application of the two statistical tests of log-likelihood and BIC (see Sect. 4.2) to compare the use of relevant adjectives in paired corpus components in different languages, reveal that differences are significant for all pair-wise comparisons except for when French tourism texts are compared to American English and Russian texts. This confirms the original observation that tourism texts in these three languages describe physical space with no statistically significant differences. Therefore, although some languages differ in terms of the emphasis they place on the description of the destination as a promotional tool, there are also important similarities. It is interesting to recall that even tourism texts in languages that were found to rely less heavily on verbs related to the reader-tourist (i.e. Russian) in the previous case study have also been found to employ fewer descriptions of physical space, while tourism texts in languages that refer more explicitly to how the tourist is to behave at a destination (i.e. British English) have also been found to favour descriptions of physical space. However, the reverse pattern has also been observed (i.e. for Greek) with a low number of verbs describing tourist behaviour but a high number of descriptions of space. This suggests that these two central elements of tourism discourse are not necessarily complementary, i.e. their distribution is not mutually dictated, and thus might vary. Further studies are needed to explore in more depth how different key elements of tourism discourse interact—one hypothesis worth testing is whether descriptions of physical space act as the backdrop for tourist activities or as standalone promotional elements (i.e. pull factors) of tourism discourse.

With the number of frequently used adjectives established, it is possible to calculate the normalised frequencies of the semantic categories to which they belong. Firstly, adjectives are divided into descriptors and classifiers, following the categorisation presented in Chapter 8. The focus in this study is on descriptors, which as can be seen from Table 9.2 is the category most adjectives belong to. The reason for this is that descriptors have a much more pronounced promotional function and therefore are more strongly associated with tourism discourse, while they can also capture the author's stance. Additionally, as the words suggest, descriptors aim to describe physical space, while classifiers aim to classify physical space with reference to something else—this can be

Table 9.2 Descriptors and classifiers in non-translated corpus components

		EN_US	EN_UK	FR	EL	RU
Adjectives (tokens)		1,587	2,637	1,791	3,812	1,892
Descriptors	Raw frequency	954	2.007	1,232	2,648	1,022
	Normalized frequency	60.11%	76.11%	68.79%	69.46%	54.02%
Classifiers	Raw frequency	633	630	559	1,164	870
	Normalized frequency	39.89%	23.89%	31.21%	30.54%	45.98%

another referent as in the case of *nearby* (nearby something else) or a specific group as in the case of *national* (specific nation). That is not to say that classifiers are not important in tourism discourse, or that they do not help us understand how people gaze at the world around them, rather that they do so to a lesser degree.

Since the focus of this exploration is on descriptors, these can be further divided into nine semantic categories following Biber et al.'s (1999) model, which has been refined based on available data from the TrAIL corpus, namely Colour, Size, Time, Beauty, Quality, Effect, Distinctiveness, Fame, and Other, with varying degrees of frequency as summarised in Table 9.3. The most prominent semantic categories are marked in bold. Normalised frequencies are calculated based on the total number of all identified adjectives in each language component which meet the threshold frequency measure of 0.18ptw and which also modify a noun denoting physical space.

The distribution of semantic categories varies considerably, with some categories, i.e. Size, Time, Beauty, Quality, and Fame, representing a high normalised frequency (i.e. above 10% in any one language) while others, i.e. Colour, Effect, Distinctiveness, and Other representing a significantly lower one, often very close to 0%. Due to their prominence, the discussion that follows focuses only on the high-frequency semantic categories. Any of the less prominent categories should not, however, be considered less important; rather, they play a secondary role in how physical space is being described—taken together they represent around 15% of all adjectives performing a descriptive function in any one language. Valid

Table 9.3 Distribution of semantic categories of descriptors in non-translated corpus components as a percentage of identified adjectives (tokens)

		EN_US	EN_UK	FR	EL	RU
Adjectives (tokens)		1,587	2,637	1,791	3,812	1,892
Colour	Raw frequency	31	36	90	50	0
	Normalized frequency	1.95%	1.37%	5.03%	1.31%	0%
Size	Raw frequency	**162**	**198**	**403**	**583**	**386**
	Normalized frequency	**10.21%**	**7.51%**	**22.50%**	**15.29%**	**20.40%**
Time	Raw frequency	**70**	**296**	**269**	**519**	**380**
	Normalized frequency	**4.41%**	**11.22%**	**15.02%**	**13.61%**	**20.08%**
Beauty	Raw frequency	**143**	**479**	**149**	**452**	**85**
	Normalized frequency	**9.01%**	**18.16%**	**8.32%**	**11.86%**	**4.49%**
Quality	Raw frequency	**229**	**369**	**0**	**119**	**0**
	Normalized frequency	**14.43%**	**13.99%**	**0%**	**3.12%**	**0%**
Effect	Raw frequency	0	68	0	237	0
	Normalized frequency	0%	2.58%	0%	6.22%	0%
Distinctiveness	Raw frequency	59	65	66	149	75
	Normalized frequency	3.72%	2.46%	3.69%	3.91%	3.96%
Fame	Raw frequency	**211**	**270**	**87**	**384**	**96**
	Normalized frequency	**13.30%**	**10.24%**	**4.86%**	**10.07%**	**5.07%**
Other	Raw frequency	49	226	168	125	0
	Normalized frequency	3.09%	8.57%	9.38%	4.07%	0%

conclusions are also easier to reach with higher frequencies. Figure 9.1 is a visual representation of the distribution of the five most prominent semantic categories across languages, according to the normalised frequencies shown in Table 9.3.

In the following sections, each of these five semantic categories will be examined in depth, both in terms of frequencies, but also, where relevant, in terms of how they are realised, i.e. which particular adjectives are used in relation to Size, Time, Beauty, Quality, and Fame. This detailed examination will reveal any cross-linguistic differences in the way in which these key components of tourism destination are conceptualised and ultimately how different societies organise and gaze at the world around them. As will be explained in subsequent sections, some of these categories can be considered 'objective' since they communicate factual information about physical space (for example, whether something is big or old), while others can be considered 'subjective' since they rely on the author's judgement or opinion (for example, whether something is beautiful or good). Therefore, an important parameter in this exploration regarding how societies perceive physical space will be the

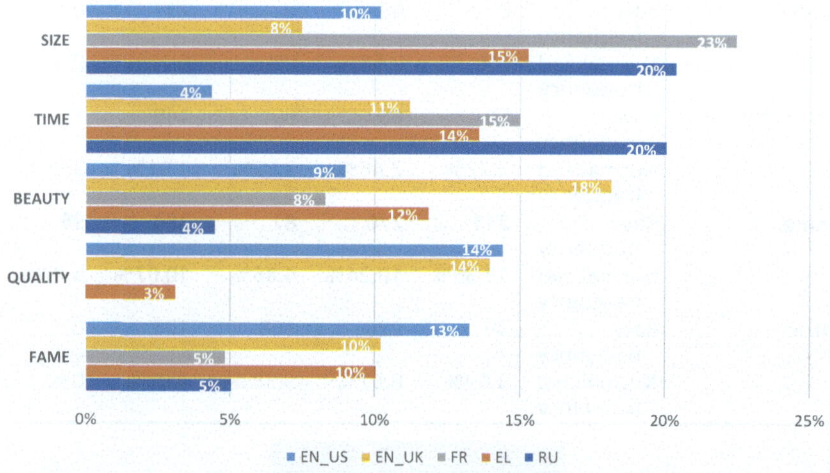

Fig. 9.1 Normalised frequencies of the most prominent semantic categories of descriptors in non-translated corpus components as % of all identified adjectives

focus on objective/subjective features. This distinction between objective and subjective semantic categories is naturally relative; the choice of a specific adjective is always a rather subjective choice. Compared to each other, it is, however, possible to argue that some semantic categories exhibit more elements of objectivity/subjectivity than others. Finally, the aim here is not to argue about whether, for example, a comment about the size of a building corresponds to reality (e.g. if it is indeed big), but rather how the author aims to present such information, that is, focusing on what could be considered objective features rather than subjective perceptions of physical space.

9.2 Size

The semantic category of Size, which refers to the dimensions and extent of space, can be considered an objective one as it refers to something that can be easily perceived through the eyes, for example, whether a park is large or a building is tall. Distribution of this semantic category across languages varies greatly with French and Russian tourism texts showing a strong preference for its use (22.50% and 20.40% respectively), since its frequency is higher in these two language components than in the others, and English tourism texts relying on it only occasionally (10.21% in American English and 7.51% in British English). The middle space is occupied by Greek tourism texts (15.29%). The statistical tests, conducted on the raw frequencies, reveal that these differences are significant in all cases apart from American English versus British English and French versus Russian tourism texts. This confirms a strong preference towards Size adjectives in Russian and French. It also confirms a relatively lower preference in English, irrespective of the variety.

Overall, it appears that there is consensus across the languages examined here that largeness is an aspect of physical space that makes it attractive to prospective tourists, while there also appears to be a hierarchy regarding which aspects of Size are prioritised. First comes dimension with both *big* and its opposite *small* and their equivalents in other languages, as illustrated in Examples 9.1 and 9.2—if physical space is to be described based on its size the first distinguishing criterion is whether it is big or small. It is worth noting that *petit* (*small*) in French has a particularly positive semantic prosody which is not typically found in English, something that would further explain the preference for this adjective in French tourism discourse. For example, compared to *small* in British English tourism texts, *petit* (*small*) is used approximately four times more frequently in French ones.

Example 9.1

Below are some of London's ***biggest*** *and best parks.*
Source www.visitlondon.com

Example 9.2

ST *Elle est lovée le long du plus* ***petit*** *fleuve de France, la Veules, long de 1149 mètres!*
BT *It is nestled along the* ***smallest*** *river in France, the Veules, 1149 meters long!*
Source www.france.fr

Second comes height, which is realised only through *high* and its equivalents, as illustrated in Example 9.3, but not through its opposite *low*. It is interesting that although smallness can be considered attractive in tourism, shortness is not.

Example 9.3

ST *Ο Όθος, το* ***ψηλότερο*** *χωριό του νησιού, χτισμένο αμφιθεατρικά σε υψόμετρο 510 μ, συχνά μοιάζει να χάνεται μέσα στα σύννεφα!*
BT *Othos, the* ***highest*** *village of the island, built amphitheatrically at an altitude of 510 m, often seems to disappear in the clouds!*
Source www.visitgreece.gr

Next come other aspects, such as extent or shape, for example, *vaste* (*vast*) and *просторный* (*spacious*) in French and Russian, respectively (which can, however, be considered as closely related to *big*), as illustrated in Example 9.4, and *στενός* (*narrow*) in Greek, as is illustrated in Example 9.5. Once again, there are similarities between French and

Russian tourism texts: in both languages the category of Size is realised through adjectives focusing on the same elements of physical space, i.e. spaciousness.

Example 9.4

ST *К культурному центру прилегает* ***просторный*** *сад, огибающий асимметричное здание, – летом тут проводят большие музыкальные фестивали.*
BT *The cultural center is adjoined by a* ***spacious*** *garden that wraps around the asymmetrical building and hosts large music festivals in the summer.*
Source www.discovermoscow.com

Example 9.5

ST ***Στενά*** *δρομάκια, αμέτρητα διώροφα σπίτια με κεραμιδένιες στέγες, όλα διαφορετικά, όλα όμορφα.*
BT ***Narrow*** *streets, countless two-story houses with tiled roofs, all different, all beautiful.*
Source www.discovergreece.com

Overall, the semantic category of Size appears to be given very different attention by some societies, such as the Russian and the French one, classifying the world around them to a large extent based on size considerations, others like the English ones considerably less so. Additionally, the more emphasis there is on Size, the more adjectives will be used from the lower levels of the hierarchy (dimension-height-shape/extent). Specifically, Russian, French, and Greek texts employ the greatest range of adjectives, English texts the smallest. Based on these results, it appears that the tourist gaze in official English tourism websites does not prioritise aspects of size, extent, or shape as much as other languages do, that is they rely less on this objective category to classify physical space.

9.3 Time

A prominent semantic category is that of Time, which in the present case study refers to the chronology or age of physical space. Like Size, it can be considered a relatively objective category and an aspect of physical space that can be easily perceived, especially since many destinations boast a long history. A preference towards Time adjectives is found in Russian (20.08%) and French (15.02%), followed by Greek (13.61%) and British English (11.22%) tourism texts, while normalised frequencies of this semantic category are markedly lower in American English (4.41%). Differences are statistically significant across all pairwise comparisons, according to tests conducted on raw frequencies, the only exceptions being the comparison of Greek tourism texts with those in British English and French. Interestingly, there is also a statistically significant difference between American and British English tourism texts. This confirms that there is a lot of variation here. However, neither English tourism texts, nor American English ones in particular, appear to prioritise time criteria compared to the other languages examined here, while a much stronger preference for this objective category is once again found in Russian texts.

Regarding specific lexical choices, there is an interesting observation in relation to the distribution of old/new. In all languages, apart from Russian, *old* and *ancient* and their equivalents are the most frequently used adjectives in this semantic category, as illustrated in Examples 9.6 and 9.7, where the emphasis on aspects of the destination that can be characterised as old is very prominent. In fact, *old* is the only Time adjective used in American English tourism texts, while French texts do not employ any adjective related to newness (such as *contemporain*), with only *ancien* (*ancient*), *vieil* (*old*), and *antique* (*ancient*) being found.

Example 9.6

History Quest Amelia Island is one of the state's most historic locations, claiming many of Florida's "oldest": ***oldest*** *bar,* ***oldest*** *inn,* ***oldest*** *lighthouse and oldest newspaper.*

Source www.visittheusa.com

Example 9.7

ST *De nombreux vestiges du passé subsistent: **l'ancienne** prison avec le cachot de Cyparis, **l'ancienne** église du Fort, l'Habitation Levassor, **l'ancien** théâtre...*

BT *Numerous relics of the past remain: the **old** prison with the Cyparis dungeon, the **old** church of the Fort, the Levassor dwelling, the **old** theater...*

Source www.france.fr

However, in Russian, the most frequent Time adjective is *новый* (*new*), as illustrated in Example 9.8. Related to this point, in languages where both *old* and *new* have been found (i.e. British English, Greek, and Russian), the occurrences of *old* and its synonyms *ancient*, *former*, and *traditional* and their equivalents by far outnumber those for *new* and *modern* and their equivalents in all languages apart from Russian where their distribution is more balanced (221 adjectives related to *old* and 159 related to *new*).

Example 9.8

ST *Со стороны Васильевского острова была значительно расширена набережная, здесь появилась **новая площадь** – площадь Трезини.*

BT *On the Vasilievsky island side, the embankment was greatly extended and a **new square**, Trezzini Square, was created.*

Source www.visit-petersburg.ru

What can be concluded from this examination of the distribution of old/new is that in most societies, oldness is an aspect of the destination singled out and identified as potentially attractive. This is not particularly surprising considering that history is a central promotional pull factor in the tourism industry, while there is also historical tourism with the primary purpose of exploring the history and heritage of a place. Tourism is also for many people a means of escaping from the modern, often hectic, life, and exploring the past is promoted as a good alternative, as manifested in Example 9.9.

Example 9.9

Follow the country's movie trails in the footsteps of your favourite characters and discover Scottish film locations that include vast beaches, ***ancient castles****, mountains, idyllic rolling hills, distinctive cities and historic towns and villages.*
Source www.visitscotland.com

What is, however, surprising is that this pattern is not observed across all languages examined here, with Russian (once again) clearly standing out from the rest. The emphasis on newness and modernity when describing physical space in Russian tourism texts, which can be clearly seen in Example 9.8, might be associated with the desire to promote itself as a modern country and destination on a par with Western countries. This self-image becomes part of the tourist gaze, which then focuses on elements of physical space that reflect this self-identity further reinforcing it in a process that feeds into itself.

The question arises whether visitors to Russia would be equally interested in the modern aspects of the destination or whether they would prefer to explore its past. Similarly, it is possible that Russian visitors to other countries might not prioritise their modern aspects. This is an area where country branding and destination image merge and it is difficult to separate one from the other. For the purposes of this case study, it will be assumed that non-Russian-speaking tourists are more used to focusing on the old rather than the new, while Russian-speaking tourists may be accustomed to using both old and new when it comes to descriptions of physical space. However, this is highlighted as a fruitful avenue of further investigation.

Overall, the examination of the semantic category of Time suggests that a clear distinction can be made regarding how this category is treated in English tourism texts and those in other languages, with English texts, particularly American, directing the tourist gaze less frequently towards the age and chronology of physical space. Russian tourism texts stand out even more from the rest with the equal emphasis they place on both old and new. This is another objective category, after Size, where difference can be observed cross-linguistically, with English tourism texts once again paying less attention to this category than other languages. As noted, these differences might also be influenced by country branding, adding an additional layer of consideration during translation.

9.4 Beauty

While the previous two semantic categories can be characterised as being relatively objective, the Beauty category is perhaps one of the most subjective ones, because it refers to the aesthetics of physical space. Tourism texts often aim to create a postcard-like image of a destination by beautifying it (Francesconi 2014, p. 58) which is, at least in the verbal mode, to a large extent achieved through the use of adjectives. The focus here is not on what can be perceived as beautiful, which is inherently subjective and culture-specific, but rather on whether destinations are perceived as being beautiful across languages, therefore revealing the values of each society regarding physical space and tourism gaze.

The strongest preference towards Beauty is found in British English tourism texts (18.16%), since it is in these tourism texts that we find the highest normalised frequency for this category, followed by Greek texts (11.86%), American English (9.01%), and French texts (8.32%). The weakest occurrence is found in Russian tourism texts (4.49%). Statistical tests, conducted on the raw frequencies, reveal that differences are significant for all pair-wise comparisons apart from American English versus French and versus Greek. This is an indication that there is great variation in how this semantic category is treated across languages and the emphasis that they put on the aesthetics of a destination.

As noted, British English tourism texts appear to focus a lot on the aesthetics of physical space, and in fact approximately 20% of all adjectives frequently found in tourism texts describing Britain belong to this category. British English is also the language where the largest variety of types of Beauty adjective is employed—7 in total—further supporting the prominence of this semantic category in British tourism discourse. An example of this emphasis on the aesthetics of the destination can be found in Example 9.10. In this example, two Beauty adjectives are used in the same sentence to describe aspects of a village in Wales.

Example 9.10

Portmeirion, the unique Italianate village created in the late twentieth century, is a must-visit with its ***picturesque*** *buildings and* ***beautiful*** *sub-tropical gardens.*
Source www.visitbritain.com

British English tourism texts are also not afraid to rely on clichés when it comes to describing how beautiful a destination is, as seen in Example 9.11 where the phrase *chocolate-box pretty* is being used to describe Cotswold villages. This phrase has connotations of boredom or conventionality, according to the Collins dictionary (Collins online) (perhaps with the meaning of traditional here) which might not appear appropriate when trying to promote a destination for certain tourist groups at least. Yet, it is assumed that British tourism discourse tolerates such stereotypical descriptions.

Example 9.11

With this itinerary, tour the historic towns and ***chocolate-box pretty*** *villages of the Cotswolds region of south west England.*
Source www.visitbritain.com

A much smaller number of types of Beauty adjectives, i.e. only 2, is found in French and Russian tourism texts, corresponding to 7% and 6% of all adjective types in this semantic category in each language respectively.[1] In French in particular, the structure *l'un(e) de plus beaux/belles* (*one of the most beautiful*) is employed irrespective of what type of physical space is being described as seen in Examples 9.12 and 9.13. This suggests that despite the low occurrence of such adjectives in French tourism discourse, beauty is used as a rather stereotypical way of describing physical space, with standard expressions associated with it.

[1] Please note that no table is provided with specific details of this, due to the amount of data.

Example 9.12

ST *À Eguisheim, **l'un des plus beaux villages** de France, on peaufine ses connaissances sur l''architecture médiévale.*
BT *In Eguisheim, **one of the most beautiful villages** in France, you can learn more about medieval architecture.*
Source www.france.fr

Example 9.13

ST *On la considère souvent la plage de La Salis comme **l'une des plus belles plages** de la Côte d'Azur.*
BT *La Salis beach is often considered **one of the most beautiful beaches** on the Côte d'Azur.*
Source www.provence-alpes-cotedazur.com

Finally, American English and Greek employ 3 and 5 types of Beauty adjectives respectively, which corresponds to 11% and 9% of all adjective types in this semantic category in each language respectively. This demonstrates that although it appears that British English shows the strongest preference towards categorising physical space through the lens of beauty based on the total number of all Beauty adjectives, a rather strong preference is also observed in American English based on the number of adjective types.

A similarity between Greek and Russian is that in both languages, the most frequently used Beauty adjective is the equivalent of *picturesque* as illustrated in Examples 9.14 and 9.15, with the equivalent of *beautiful* ranking second.

Example 9.14

ST *Οι διακοπές σας στη Ρόδο "απαιτούν" σίγουρα μια επίσκεψη στη **γραφική** αυτή πόλη του νησιού.*
BT *Your holidays in Rhodes "demand" a visit to this **picturesque** town of the island.*
Source www.discovergreece.com

Example 9.15

ST *В Сосновке расположено одно из самых* ***живописных*** *мест черноморского побережья.*
BT *In Sosnovka is located one of the most* ***picturesque*** *places of the Black Sea coast.*
Source www.russia.travel

There is an emphasis here on beauty derived from physical spaces that are considered old-fashioned or traditional (similar to the *chocolate-box pretty villages* in British English). French and American English, on the other hand, do not appear to employ *picturesque*. This emphasis on oldness as an attractive feature of physical space can be linked to the results obtained from the analysis of Time adjectives. A generally higher frequency of Time adjectives was also found in Russian, Greek, and to some extent British English, while a very low one was observed in American English. The only exception to this pattern is French tourism discourse, where although *pittoresque* (*pisturesque*) was not noted, there is a strong preference towards Time adjectives, although interestingly enough not for *traditionnel* (*traditional*). Conversely, *traditional* and *παραδοσιακός* (*traditional*) are frequently used in British English and Greek tourism discourse, but not in Russian. While the picture is certainly complex, it reveals that there are interconnections between the different semantic categories, which further supports the idea that different aspects of physical space are prioritised by different societies not in a random fashion but in a meaningful way.

Overall, the examination of the semantic category of Beauty suggests that there is a lot of variation in this category across languages, with British English tourism texts interpreting physical space as beautiful more often than tourism texts in other languages. Unlike the more objective semantic categories examined so far, which English tourism texts appeared not to prioritise, the opposite is true for the more subjective category of Beauty. Connections can also be made between Beauty and Time, and the emphasis that societies place on age as a distinguishing feature of physical space, creating a rather challenging and complex translation situation.

9.5 Quality

Closely linked to Beauty is the semantic category of Quality, which refers to how good physical space is. This is another subjective category in the sense that different people are likely to rank different destinations and even different elements of the same destination differently. There is also a competitive element in this category through the implication that some aspects of physical space might be better than others and therefore worth visiting.

The strongest preference towards Quality is found in American English tourism texts (14.43%), since it is in this corpus component that it is found most frequently, closely followed by British English ones (13.99%). A low frequency is found in Greek tourism texts (3.12%) while this category appears to not be frequently employed in French or Russian tourism texts (0%). Statistical tests, conducted on the raw frequencies, reveal that differences are significant only when Greek texts are compared to English ones, irrespective of the language variety (no tests have been conducted for languages with 0% frequency). In other words, both American and British English tourism texts employ Quality adjectives to describe physical space with a similar frequency, significantly more frequently than Greek tourism texts. Quality appears to be a very distinguishing characteristic of English tourism discourse.

As with the Beauty category, the largest variety in terms of adjectives is noted in British English (5 types in total, corresponding to 11% of all adjective types in this semantic category for this language component), and American English (4 in total, corresponding to 17% of all adjective types). Greek tourism texts use 2 adjective types, corresponding to 4% of all adjective types. These quantitative findings further support the general prevalence of Quality adjectives in English in comparison to both the other semantic categories in the English-language tourism texts and the normalised frequency of this category in other languages.

In terms of specific lexical choices, the most commonly used adjective in both varieties of English is *good* (Example 9.16), while in Greek it is ιδανικός (*ideal*) (Example 9.17), which is also found in British English, with καλός (*good*) in second place. Both varieties of English also employ *top*, *great*, and *perfect*. It is worth commenting on the fact that with the exception of *good* and its equivalent in Greek, all other adjectives in this category are non-gradable, because they already encode the concept of superiority. A closer inspection of *good* and καλός (*good*) reveals that the most frequent form of this adjective is *best* in English and καλύτερος (*best*) in Greek. Therefore, the focus is not simply on quality but on the highest possible quality.

Example 9.16

From acres of flowers to historic Japanese gardens and lush hidden gems, read on and discover the ***best*** *gardens in L.A.*

Source www.discoverlosangeles.com

Example 9.17

ST *Οι μικρές νησίδες του Παγασητικού αποτελούν* ***ιδανικούς*** *ψαρότοπους και αποτελούν σημεία συνάντησης των ψαροντουφεκάδων.*

BT *The small islands of Pagasitikos are* ***ideal*** *fishing grounds and are meeting points for snorkelers.*

Source www.visitgreece.gr

Overall, striking differences are noted for the semantic category of Quality, and it can be considered as a distinguishing characteristic of English tourism discourse, which places emphasis on how entire destinations can be compared to others. Other languages either do not categorise physical space in terms of how good it is, or do so very occasionally, i.e. Greek tourism texts. This is another subjective category, after Beauty, where differences can be observed cross-linguistically, with English tourism texts once again paying more attention to this category than other languages.

9.6 Fame

The final semantic category to be examined in this chapter is Fame, which refers to how liked or approved physical space is considered to be typically by other visitors to the destination. Tourism as an activity often focuses on the most popular landmarks, which would justify why this category is particularly prominent. Similarly, a typical strategy for attracting visitors to a place is by creating a sense of must-see, which is tightly linked to the idea of how iconic a destination or specific aspects of it are. If many people enjoy visiting a specific physical place, then the reader must do so too so as not to miss out. What is also worth mentioning is that Fame can be considered as having elements of both an objective and a subjective category: while it is possible to identify popular places or landmarks based on how many visitors they attract, what is perceived and promoted as famous in tourism discourse is also open to interpretation and might be considered the author's own judgement. In that sense, Fame is understood here in a rather loose way and not tightly linked to the specific notion of popularity alone.

The strongest preference towards Fame adjectives is observed in American English tourism texts (13.30%), since it is in these texts that these are found most frequently, followed by British English ones (10.24%) and Greek (10.07%). The weakest preference, that is where they are found least frequently, is observed in Russian (5.07%) and French tourism texts (4.86%). Statistical tests, conducted on the raw frequencies, reveal that differences are significant across all pair-wise comparisons except for American English versus British English, American English versus Greek, British English versus Greek, and French versus Russian. This confirms the observation that Fame is assigned different importance across some languages, and that that there is a clear distinction between, on the one hand, English (whether American or British) and Greek, and on the other hand, French and Russian. This is not the first time that tourism discourse in French and Russian appears to behave similarly. In English tourism texts, both American and British, typical Fame adjectives include *popular, famous,* and *iconic*, as illustrated in Example 9.18 where two of these adjectives are used in the same sentence providing evidence of the preference that American English tourism texts show towards this semantic category.

Example 9.18

*South Beach – from South Pointe Park north to 23rd Street – is the most **famous** of the trio, home to sandy strands, swaying palms and the pastel-colored Art Deco buildings that have become synonymous with this **iconic destination** and render it so undeniably photogenic.*

Source www.miamiandbeaches.com

The adjective *iconic* constitutes an interesting choice, only found in American and British English texts, since it can be considered as the most subjective of the Fame adjectives, implying a sense of admiration and respect and therefore a personal assessment by the author. This might be evidence of some subjectivity associated with the Fame category in both varieties of English. In quantitative terms, British English shows a stronger preference towards the use of the adjective *iconic*, which accounts for approximately 30% of all Fame adjectives, compared to 18% in American English. Overall, the concept of Fame is released linguistically in the same way through similar linguistic choices, and emphasis in both varieties of English, with a slightly stronger preference towards more subjective choices in British English (i.e. as manifested through the use of *iconic*).

Interestingly, Greek tourism texts employ the largest variety of types of Fame adjectives (6 in total), all of which are equivalents of *popular* in English, with the exception of *σημαντικός* (significant), which is also the most commonly used Fame adjective in Greek tourism discourse (Example 9.19), and *σπουδαίος* (*important*). This specific emphasis on 'significance' is noteworthy and appears to be an effort to shift attention from the notion of popularity (in the sense of approval and likeability) which is based on what other visitors of the destination might think towards the idea of importance which relies a lot more on the author's judgement, thus making it a subjective adjective. This further establishes the author (or the destination promoter more accurately) as an expert whose opinion needs to be trusted. Overall, approximately 40% of all Fame adjectives in Greek tourism texts can be considered as subjective.

Example 9.19

ST *Ακόμα νοτιότερα θα βρείτε το, επίσης* ***σημαντικό*** *σπήλαιο Κολώνες.*
BT *Even further south you will find the also* ***important*** *Kolones cave.*
Source www.discovergreece.com

Regarding French and Russian tourism discourse, only two types of Fame adjectives have been observed in each language. Russian uses equivalents of the adjective *popular* (i.e. *известный* and *популярный*), as illustrated in Example 9.20, while French makes use of the adjectives *célèbre* (*famous*) and also *mythique* (*legendary*) which intensifies the significance of physical space, almost exaggerating it (Example 9.21). It appears that Russian focuses very clearly on the notion of popularity, which can be considered objective, while French assigns almost equal importance to both the subjective and objective realisations of this category—the two French adjectives are used with almost the same frequency.

Example 9.20

ST *Всемирно* ***известный*** *город-музей интересен комплексом Спасо-Евфимьевого монастыря XIV века с роскошным Спасо-Преображенским собором.*
BT *The* ***world-famous*** *city-museum is interesting with the complex of the Spaso-Evfimiev Monastery of the XIV century with the magnificent Spaso-Preobrazhensky Cathedral.*
Source www.russia.travel

Example 9.21

ST *Rien d'étonnant à ce que* ***l'avenue mythique*** *qui défile sous vos yeux soit la première vision de Paris qui s'offre à Andy (Anne Hathaway) et Miranda Priestly (Meryl Streep), à travers les vitres fumées de leur taxi.*
BT *It's no surprise that* ***the legendary avenue*** *that passes before your eyes is the first glimpse of Paris that Andy (Anne Hathaway) and Miranda Priestly (Meryl Streep) get through the smoked windows of their taxi.*
Source www.paris.info

Overall, there are significant differences in terms of how Fame adjectives are treated across some of the languages examined here. There is a clear preference towards this semantic category in English and Greek tourism discourse and a significantly lower one in French and Russian. Differences involve not only the frequency with which such adjectives are found in tourism texts, but also how exactly they are realised linguistically. Perhaps the most interesting finding for this category involves the focus that different languages place on the more objective/subjective adjectives through which this category can be realised, with Greek tourism texts exhibiting a tendency to tone down the objective elements and present the author as an expert, a pattern also found in English tourism texts, particularly British ones, while French and Russian tourism texts show a tendency towards the opposite. Once again, this manifests how semantic categories are linked to each other and how individual societies make systematic choices when it comes to classifying physical space.

9.7 Conclusions

The examination of the most prominent semantic categories of adjectives, and more specifically descriptors, modifying physical space has helped shed light on how the different languages examined as part of the TrAIL corpus classify and interpret elements of physical space, that is, how the tourist gaze is realised in these languages. Although not a distinct part of Biber et al.'s (1999) framework used for the analysis of data, it has been possible to group these prominent semantic categories of descriptors into objective, subjective, or with elements of both. Specifically, Size and Time belong to the objective group, Beauty and Quality to the subjective one (corresponding to Biber et al.'s Evaluative category), while Fame exhibits elements of both. It is worth recalling that these categories (i.e. objective vs subjective) are examined here in relative terms: instead, the focus is on the impression obtained from the tourism text, rather than on whether it represents an accurate depiction of reality (e.g. whether a building is actually old or not and based on particular criteria).

Table 9.4 summarises the distribution of these adjectives according to the classification of the relevant semantic category as 'subjective' or 'objective' across the languages examined here. The labels 'very',

Table 9.4 Distribution of objective/subjective descriptive adjectives in each language component of the TrAIL corpus

	American English	British English	French	Greek	Russian
Objective	Medium	Medium	Very High	High	Very high
Subjective	High	Very high	Low	Medium	Low
Both	Medium	Medium	Low	Medium	Low

'high', and 'low' are based on the distribution of semantic categories of descriptors in non-translated corpus components (see Table 9.3). Size and Time adjectives are regarded as 'objective' categories, Beauty and Quality as 'subjective', and Fame has been found to include elements of both a 'subjective' and 'objective' category. Their percentages in each language component are then combined to generate Table 9.4. A very high frequency refers to frequencies above 30% of all adjectives, low to anything below 10%. Frequencies between 10% and 20% are considered medium, and those between 20% and 30% are considered high.

From the table, it is possible to identify three distinct patterns. First, English language tourism texts (whether American or British) show a strong preference towards subjective descriptions, and a medium one towards objective ones. Similarities are also observed in the semantic category that have elements of both subjectivity and objectivity, i.e. Fame. Having said that, the detailed analysis also reveals that British English texts emphasise more the subjectivity associated with Fame, resulting potentially in a slightly different effect. In English tourism texts, it seems that the world is interpreted through a mixture of personal evaluations and factual characterisation, with a preference towards the former. It can be argued that in American and British societies, more importance is placed on the individual who is experiencing physical space (and how they perceive it, an inherently subjective process), rather on the physical space and how it exists independently of the observer. For example, a mountain can be high without the need for anyone to confirm this, but it can be beautiful only through the eyes of an observer. Therefore, the producers of these texts establish themselves as experts whose views need to be trusted. Meanwhile, readers of these texts are expected to trust these views, despite the subjectivity associated with them. This point is particularly pertinent in tourism, since, unlike other expensive products, it cannot be trialled or sampled and therefore potential customers can only rely on descriptions. The fact that they can presumably be convinced through more subjective descriptions reveals an interesting aspect of English-speaking societies. This aspect can be associated with the tourism phase these societies are experiencing. Although more research is required, it is possible that the post-sightseeing phase can also be associated with more risk-taking, in the sense that the actual experience of the destination is more likely to differ compared to a subjective rather an objective description of it.

The second pattern identified is found in French and Russian tourism texts, where there is a very clear preference towards objective descriptions over any other that might involve subjectivity. In French and Russian tourism texts, it seems that the world is primarily organised around factual information, avoiding any personal evaluations. Hence, more importance appears to be given to the physical space rather than how visitors might perceive or experience it. Credibility is established in these cases through these objective descriptions which can be considered reliable, the view of the author is therefore of secondary importance. This suggests that French and Russian speakers might avoid risk-taking and are likely to be convinced to visit a destination through more objective descriptions. It is interesting that these two societies have been found, in the previous case study, to belong to slightly different phases of tourism; with the Russian society clearly still in the sightseeing phase and the French one closer to the post-sightseeing phase but not having reached it yet.

The final pattern is that observed regarding Greek tourism texts, where although there is a preference for objective descriptions, there is also a (medium) tendency to rely on subjective ones. This positions Greek tourism texts between English ones on the one hand and French/Russian ones on the other. It also suggests that Greek-speakers require credibility to be established through seemingly factual information, but that they also tolerate personal evaluations by the text producers, which might act as a pull factor. In other words, while priority is given to the physical destination, some attention is also paid to the individuals who experience it. Linking this back to the phase of tourism, the Greek society has been found to exhibit elements of both the sightseeing and the post-sightseeing phase, not fully having reached the latter yet, which appears to be further supported by the distribution of adjectives.

Although it has not been an aim of the second case study, it has been hard to ignore the links between the distribution of adjectives and the tourism phase that a society is experiencing. And yet, these connections appear logical. For example, it is not surprising that the post-sightseeing phase would be associated with more subjectivity and more emphasis on how an individual might perceive a destination, while the sightseeing phase might favour more 'traditional' descriptions focusing more on information value. This lends further credibility to the idea of tourism phase, a rather under-studied concept, and suggests that it can manifest itself through a wide range of different linguistic means (only some of which have been examined in this book) which are linked to all aspects of tourism promotion, that is both in terms of things to do and what to see. More studies of this sort are required to better understand the importance of the tourism phase and how it can be realised linguistically.

Findings from this case study regarding how physical space is perceived across languages, can inform translation, and shape the amount and type of adaptation required when tourism texts are translated into other languages. For example, if translation is into English, as is often the case, adaptation might involve the introduction of more subjective descriptions associated with Beauty and Quality, while the reverse could be expected when translation is out of English. Lack of adaptation, by employing the same semantic categories of adjectives used in the source text in the target text, is likely to have a negative effect on the success of tourism promotion. A tourism text that describes a destination to an audience who is still experiencing the sightseeing phase using a high frequency of subjective adjectives is likely to create a sense of insecurity in the readers and discourage them from visiting that destination. Similarly, a text which aims to reflect the post-sightseeing phase and describes a destination using mainly objective adjectives is likely to make it unappealing. The next chapter will examine whether translators manage to recognise in which cases adaptation is required regarding descriptions of physical space and the form it needs to take in order to produce texts that reflect how the target audience is most likely to gaze at the world around them.

References

Biber, D., S. Johansson, G. Leech, and S. Conrad. 1999. *Longman Grammar of Spoken and Written English*. Longman.

Chocolate-Box Definition and Meaning|Collins English Dictionary, www.collinsdictionary.com/dictionary/english/chocolate-box. Accessed 15 March 2024.

Francesconi, S. 2014. *Reading Tourism Texts: A Multimodal Analysis*. Channel View Publications.

10

Translating Physical Space

The previous chapter explored how space is conceptualised and promoted in different languages examined as part of the TrAIL corpus, and how this could be potentially linked to the tourism phase each society is currently experiencing. Notably, a difference was observed regarding the preference for describing physical space using what were classified as objective versus subjective descriptive adjectives. Following on from this, this chapter will examine how these differences are negotiated in translation, and whether and to what extent the translated texts succeed in describing the destination in a way that encodes how different societies, notably that of the target reader-tourist, gaze at physical space and which elements of it they appear to prioritise. A successful translation in that regard is also likely to make the destination more appealing to the target reader and increase the chances of their visiting it. As with Chapter 7, the discussion is organised around translated languages: English, French, and Russian. In each section, comparisons are made between target texts and both their source texts and non-translated texts in the target language.

Adaptation in this chapter is examined in relation to the distribution of the semantic categories of adjectives and, where relevant, the specific

S. Malamatidou, *Translating Tourism*, Palgrave Studies in Translating and Interpreting, https://doi.org/10.1007/978-3-031-49349-2_10

adjectives used in each category. The focus is on the five most prominent categories identified in the previous chapter, that is, Size, Time, Beauty, Quality, and Fame. Translations that come close to patterns found in respective non-translated texts, either through adaptation or through the recognition that adaptation is not required, are regarded as successful in negotiating the difference between source and target society as they (re)create an image of tourism that is in accordance with how target readers are thought to perceive the world, while at the same time maximising the promotional effect of the text.

It is worth recalling that adaptation is understood here—as also in Chapter 7—as changes that lead to a better fit and go beyond the meaning of words (see Chapter 3 for a more detailed discussion). To recap, the degree and type of adaptation is established based on statistical significance tests, as explained in Table 4.4. For example, whenever successful adaptation is mentioned that means that the following two criteria have been met. The first is a comparison of the non-translated texts in the relevant two languages that shows them to be different at a statistically significant level, indicating the need for adaptation in translation. The second comparison is between the non-translated texts now serving as source texts and their target texts; if adaptation has been successful, a statistically significant difference will also be noted here, while no statistically significant difference will be noted between the target texts and the non-translated texts in that language. For ease of reference, and to avoid a number-heavy reporting of findings, in this chapter, only normalised frequencies of the adjectives assigned to the various semantic categories will be reported, and results of statistical significance tests, provided by the online calculator, are implied in the adaptation categories reported.

10.1 Tourism Texts Translated into English

Similar to the previous case study, no distinction is being made here between British and American English for translated texts, since the variety of English is typically not specified on the tourism websites examined as part of the TrAIL corpus, where an international version of English appears to be used. This decision is further supported by the fact that no statistically significant differences have generally been identified between American and British English when comparing non-translated texts. Generally speaking, whatever adaptation patterns have been observed for one variety have also been observed for the other.

The first step before comparing the frequency distribution of semantic categories is to establish the overall frequency of all (i.e. both descriptors and classifiers) adjectives which meet the threshold frequency measure of 0.18ptw and which also modify nouns denoting physical space in non-translated texts, and in translated texts and their source texts, i.e. the non-translated texts in the relevant languages (Table 10.1). It is worth noting that, with the possible exception of translations from Russian, target texts appear to stay very close to their source texts. It will be possible to gain a more detailed picture when individual categories are examined in detail.

The next table (Table 10.2) presents the distribution of the most prominent semantic categories of descriptive adjectives in order to understand adaptation patterns. The total number of adjectives refers to the adjective tokens which meet the frequency threshold criterion of 0.18 ptw and which modify nouns denoting physical space. This number includes both descriptors and classifiers. The first row for each semantic category reports the raw frequency, the second the normalised

Table 10.1 Normalised frequencies of most frequent adjectives modifying nouns denoting physical space in non-translated English, source texts (i.e. non-translated texts in French, Greek, Russian), and translated English corpus components

		Tokens	Adjectives (types)	Adjectives (tokens)	
				f	nf (ptw)
Non-translated texts	*EN_US*	170,520	27	1,587	9.30
	EN_UK	170,704	47	2,637	15,44
Source texts	*FR*	170,060	29	1,791	10.53
	EL	170,601	57	3,812	22.34
	RU	170,881	34	1,892	11.07
Translated texts	*FR-EN*	149,421	32	1,714	11.47
	EL-EN	170,369	48	3,510	20.60
	RU-EN	186,169	21	1,286	6.90

frequency calculated out of all identified adjective tokens in that corpus component. The remaining adjectives that fall outside the five semantic categories are not included in the table.

Table 10.3 summarises adaptation patterns. A detailed presentation of what each type of adaptation refers to can be found in Chapter 4 (Table 4.4). As perhaps expected, varied patterns are noted even within the same language pair.

Although high degrees of adaptation are generally required when tourism texts are translated into English, due to the differences between non-translated English tourism texts and those in the other languages examined here, as evidenced in Chapter 9, there are only a few instances of successful or weak adaptation. There is also one case where adaptation is in the opposite direction and numerous instances of no adaptation. Overall, different language pairs appear to follow different patterns, with French-English making the strongest attempt to adapt, and Greek-English and Russian-English making almost no attempt towards adaptation. To examine semantic categories in more depth and to identify what effect the different degrees and types of adaptation have on the discourse of tourism, the discussion will now focus on individual language pairs.

10.1.1 Translations into English from French

When French tourism texts are translated into English, a generally high degree of adaptation might be expected as a result of the differences between French and English (both American and British) non-translated tourism texts established during the previous stage of analysis. When French tourism texts are translated into English, overall a medium rather than high degree of adaptation is noted. Successful adaptation is only noted with Fame adjectives, while no adaptation is found regarding Size and Time adjectives. A weak attempt towards adaptation is observed in the case of Beauty and Quality adjectives. Figure 10.1 summarises the quantitative findings presented earlier in Table 10.2.

Following the same principles as those outlined in Chapter 7 for verbs, in this and all similar figures in the rest of this chapter only the five most prominent semantic categories of descriptive adjectives, identified

Table 10.2 Normalised frequencies of most prominent semantic categories of descriptive adjectives in non-translated English, source texts (i.e. non-translated texts in French, Greek, Russian), and translated English corpus components as a percentage of identified verbs (tokens)

		NT TEXTS (TL)		NT TEXTS (SL)			TRANSLATED TEXTS		
		EN_US	EN_UK	FR	EL	RU	FR-EN	EL-EN	RU-EN
Total number of adjectives		1,587	2,637	1,791	3,812	1,892	1,714	3,510	1,286
Size	Raw frequency	162	198	403	583	386	379	675	234
	Normalised frequency	10.21%	7.51%	22.50%	15.29%	20.40%	22.11%	19.23%	18.20%
Time	Raw frequency	70	296	269	519	380	318	582	306
	Normalised frequency	4.41%	11.22%	15.02%	13.61%	20.08%	18.55%	16.58%	23.79%
Beauty	Raw frequency	143	479	149	452	85	224	328	51
	Normalised frequency	9.01%	18.16%	8.32%	11.86%	4.49%	13.07%	9.34%	3.97%
Quality	Raw frequency	229	369	0	119	0	179	158	0
	Normalised frequency	14.43%	13.99%	0%	3.12%	0%	10.24%	4.50%	0%
Fame	Raw frequency	211	270	87	384	96	167	376	126
	Normalised frequency	13.30%	10.24%	4.86%	10.07%	5.07%	9.74%	10.71%	9.80%

Table 10.3 Patterns of adaptation observed in translations into English (descriptive adjectives)

	FR-EN	EL-EN	RU-EN
Not Required		Beauty Fame	
Additional			
No	Size Time	Time Quality	Size Time Beauty Quality
Successful	Fame		Fame
Weak	Beauty Quality		
Over			
Reverse		Size	

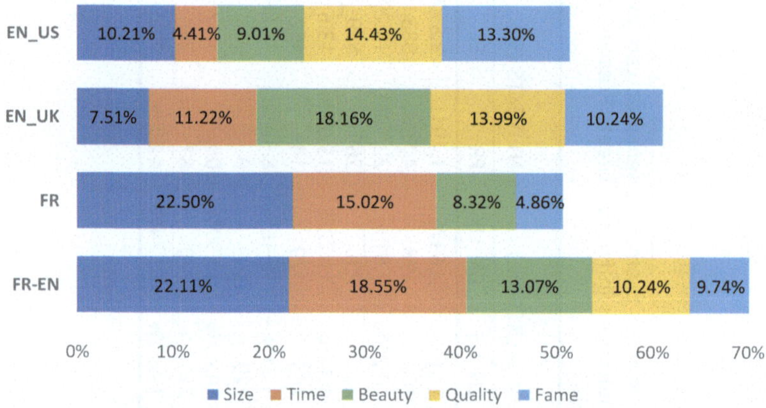

Fig. 10.1 Normalised frequencies of most prominent semantic categories of descriptive adjectives in non-translated English, French source texts, and translated English from French as % of all identified adjectives

in Chapter 9, are presented and discussed. However, their frequency is measured based on the total number of verbs identified (see Table 10.1), which explains why the total is not 100%.

Successful adaptation is observed in the case of Fame adjectives. These are less frequently used in French non-translated tourism texts (4.86%) than both American (13.30%) and British ones (10.24%), and their frequency approximately doubles in English target texts (9.74%). However, it is important to note that target texts do not completely break free from their source texts and are still influenced to a large extent by them regarding specific verb choices. Specifically, the adjective *legendary* is found in translated texts, most likely as a translation of *mythique*. The use of this adjective, which has not been found in English tourism discourse, almost exaggerates how important physical space is (Example 10.1), creating a rather unique effect.

Example 10.1

ST *Aujourd'hui, côté pile, **le mythique Saint-Germain-des-Prés** est le quartier des boutiques de luxe les plus prestigieuses qui en ont fait leur QG.*

BT *Today, **the mythical Saint-Germain-des-Prés** is home to some of the most prestigious luxury boutiques in the world.*

TT ***The legendary district of Saint-Germain-des-Prés** has a dual identity these days.*

Source www.parisinfo.com

At the same time, *iconic*, a typical Fame adjective found in both American and British tourism discourse, and which would have been a better option here, has not been found in any of the translations into English from French, further reinforcing the idea that although adaptation occurs, influences from the source text are still present and strong. However, overall a similar balance between objective/subjective descriptions, through the use of the Fame category, appears to be achieved in the English translations from French as in the non-translated English texts.

An attempt towards adaptation, albeit a weak one, is also noted in the case of Quality adjectives. This semantic category has not been observed in French non-translated tourism texts, but it is found in the English translations (10.24%), albeit with a lower frequency compared to that found in American English (14.43%) and British English tourism texts (13.99%). When a semantic category becomes prominent in translated texts, this is strong evidence of an attempt towards adaptation, and therefore even a weak attempt is considered important here. Example 10.2 illustrates how adaptation looks during translation. In this example, a significant restructuring has occurred, something that has been observed in other instances of Quality adjectives in the translations from French. This suggests that it is most probably easier for translators to move closer towards the communicative preferences of the target readers when they feel they have the freedom to adapt the source text to a significant extent, resulting in a more intuitive use of tourism language that is not directly affected by what is found in the source text.

Example 10.2

ST *Non loin de là, rejoignez le Parc national du Mercantour pour un trek 100% mobilité durable en plusieurs étapes, parcourant les Alpes de Haute Provence et les Alpes- Maritimes.*

BT *Nearby, join the Mercantour National Park for a multi-stage 100% sustainable mobility trek through the Alpes de Haute Provence and Alpes-Maritimes.*

TT *Just down the road, Mercantour National Park is* ***the perfect place*** *for a 100% sustainable mobility trek spanning the Alpes de Haute Provence and Alpes-Maritimes areas.*

Source www.france.fr

An attempt towards adaptation is also noted in the case of Beauty adjectives, but once again the attempt appears to be weak, especially compared to British English tourism texts. Specifically, French tourism texts employ Beauty adjectives with a similar frequency (8.32%) compared to American texts (9.01%) but with a much lower one compared to British tourism texts (18.16%). During translation an important increase is noted (13.07%), which brings target texts closer but not close enough to British English tourism texts, since the difference in the frequency with which Beauty adjectives are used is still statistically significant between the French translations into English and the British English non-translated texts. At the same time, this increase distances the target texts from American English tourism discourse, creating a rather unique situation. However, as this is the only case where something similar has been noted in the translations from French into English, it is not considered sufficient evidence to suggest that adaptation is mainly in the direction of British English, especially considering how weak it is. Adaptation in this case looks similar to that found in Example 10.3, where a Beauty adjective is introduced during translation.

Example 10.3

ST *En hiver, les stations de ski invitent à se lancer sur les pistes et à se ressourcer entre spas de montagne,* ***restaurants d'altitude*** *et voluptueuses adresses.*

BT *In winter, ski resorts invite you to hit the slopes and recharge your batteries in mountain spas,* ***high-altitude restaurants*** *and voluptuous addresses.*

TT *In winter, ski resorts tempt you onto the slopes then afterwards recharge your batteries at invigorating mountain spas and* ***magnificent restaurants*** *and places to stay.*

Source www.france.fr

No adaptation is noted in the case of Size and Time adjectives with reference to both American and British English. Although Size adjectives represent a lower frequency of all adjectives in both American (10.21%) and British English (7.51%) tourism texts than French texts (22.50%), target texts stay very close to their source texts (22.11%). This results in more emphasis being placed on objective descriptions of space, particularly through adjectives such as *large* and *small*. While such descriptions are typical in tourism discourse, too much emphasis on objective depictions can be at the expense of more subjective ones, such as those related to Beauty, affecting the appealing potential of the resulting tourism texts. Potentially as a result of this emphasis on Size, many more adjective types are used in the English translations compared to the non-translated texts (2 in American English, 4 in British English, and 7 in translations from French) capturing different expressions of Size, including not only *big/small* and *tall* but also *vast* and *narrow*. More interesting perhaps is the use of the adjective *little* in English translations from French. This adjective has not been found in the English tourism discourse and is a standard translation of the French *petit*. As noted in Chapter 9, this adjective has a particularly positive semantic prosody in French, often also associated with charm. This semantic prosody appears to have been transferred over to the English translations (Example 10.4), even though it is not certain to what extent this would create the same effect on the English-speaking reader-tourist. A better promotional effect, and one that is closer to how the target readers gaze at the world around them, might have been achieved, if *petit* in the example below had been translated as *charming* or *picturesque*, i.e. a Beauty adjective.

Example 10.4

ST *Faîte ensuite un crochet et montez jusqu'au* ***petit village*** *de Thoard, en passant par Digne-les-Bains, où vous pourrez admirer les champs de lavandes.*

BT *Then take a detour up to the* ***little village*** *of Thoard,* via *Digne-les-Bains, where you can admire the lavender fields.*

TT *Then make a detour and go up to the* ***little village*** *of Thoard,* via *Digne-les-Bains, where you can admire the lavender fields.*

Source www.provence-alpes-cotedazur.com

No adaptation is observed in the case of Time adjectives either. Time adjectives are more frequently used in French tourism texts (15.02%) than American (4.41%) and British ones (11.22%). During translation into English, rather than decreasing, their frequency increases (18.55%) but does not move particularly far away from that found in the source texts. Therefore, no evidence of adaptation is considered to be found here. In terms of the specific adjectives used, the 'same' ones are found in both source and target texts, with the exception of *former*. All this is evidence that no attempt has been made to adapt this category, a similar outcome to what has been noted regarding Size adjectives. Once again, this results in too much emphasis being placed on objective descriptions in the English translations from French which might have an impact on how the text is received.

Overall, the English translations of French tourism texts make some attempts towards adaptation, but these are generally limited. It is worth noting, however, that where such attempts are observed, they involve categories that can be considered subjective, i.e. Beauty and Quality adjectives, or they exhibit elements of subjectivity, i.e. Fame adjectives. Conversely, more objective categories, i.e. Size and Time adjectives, remain largely unchanged during translation. This might suggest that translators feel more reluctant to change descriptions which are perceived as objective, because that might alter the informational value of the tourism text, while subjective descriptions are perceived as more dependent on interpretation.

10.1.2 Translations into English from Greek

Although adaptation might be expected in some of the semantic categories, due to the differences between non-translated English and Greek tourism texts, as evidenced in Chapter 9, when Greek tourism texts are translated into English, no instances of successful adaptation are noted in the data. Specifically, no adaptation is noted in the case of Time and Quality adjectives, while Size adjectives are adapted in the opposite direction. Overall, as has also been observed during the previous case study focusing on verbs, a rather inconsistent approach to adaptation is noted with no clear pattern. Figure 10.2 summarises the quantitative findings presented earlier in Table 10.2.

Some adaptation is observed in the case of Size adjectives, albeit in the opposite direction. Specifically, the semantic category of Size represents a much higher frequency in the Greek tourism texts (15.29%) than in American (10.21%) and British (7.51%) texts, which suggests that the translated texts might use this category less frequently if they want to move closer to the communicative preferences of English-speaking reader-tourists. Contrary to this expectation, however, Size adjectives are

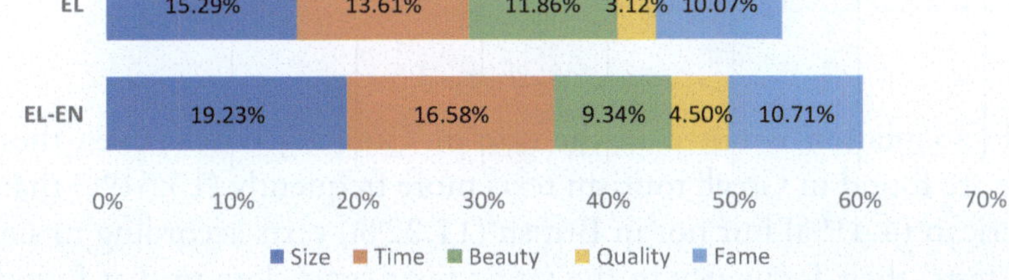

Fig. 10.2 Normalised frequencies of most prominent semantic categories of descriptive adjectives in non-translated English, Greek source texts, and translated English from Greek as % of all identified adjectives

found even more frequently in the translations from Greek (19.23%). This additional emphasis on Size is also manifested through the use of an even larger number of adjective types in the target texts than in the source texts (8 vs 5) and particularly through the use of *steep* and *long*, both of which have not been found in Greek tourism texts. Example 10.5 illustrates an interesting case of adaptation and how this might result in the over-use of Size adjectives. In this example, all key information is transferred in the translated version, but with a few additions, namely about the beach and about the Skyrian horse. When the beach is being described, a list of adjectives is used to modify it including one relating to Size. This is a case where freedom to move away from the source text does not result in successful adaptation.

Example 10.5

ST *Σκύρος: Στο ξεχωριστό αυτό νησί του Αιγαίου, μια Χώρα-καλλονή κτισμένη αμφιθεατρικά πάνω στο λόφο τους που ατενίζει το πέλαγος. Τα περίφημα σκυριανά έπιπλα και τα μπακίρια στολίζουν τα σπίτια και η μονή του Αγίου Γεωργίου με τη θαυματουργή εικόνα, επιβλέπει από ψηλά. Όσοι βρεθείτε στο διάσημο σκυριανό καρναβάλι με τους κουδουνάτους θα απολαύσετε αξέχαστες εικόνες.*

BT *Skyros: On this special island of the Aegean Sea, there is an amphitheatrically built Chora-beauty on its hill overlooking the sea. The famous Skyrian furniture and copper utensils decorate the houses and the monastery of Agios Georgios with its miraculous icon, watches over from above. Those who find themselves at the famous Skyrian carnival with the rattles will enjoy unforgettable images.*

TT *Skyros is island [sic] with a character of its own. A scenic town (Hora),* ***a long sandy main beach****, intricate wood-carved furniture, the monastery of Agios Georgios, the unique Skyrian horses and a famous carnival will set you up for an unforgettable holiday.*

Source www.discovergreece.com

No adaptation is noted in the case of Time adjectives. Even though these are found in Greek tourism texts more frequently (13.61%) than in American (4.41%) but not in British (11.22%) texts according to statistical tests, their frequency in the target texts stays close to that found in source texts (16.58%) but increases slightly moving even further away from the frequencies found in both American and British English non-translated tourism texts. This results in more emphasis being placed on objective descriptions of space, particularly through the use of adjectives such as *ancient*. This is to some extent understandable and even expected, as Greece is (stereo)typically associated with antiquity for tourism promotion purposes. However, a very high use of *old* and *traditional* is also observed in the target texts, which is more directly related to how physical space is classified, rather than a particular unique aspect of the destination. Even adjectives focusing on newness, such as *new* and *modern* are more frequently found in the translations from Greek than the comparable English non-translated text. This implies that there is indeed no attempt towards adaptation but rather an effort to promote physical space in a particular way in accordance with the unique features of the destination and not the expectations of the English-speaking reader-tourist.

Similar to Time adjectives, no adaptation is observed in the case of Quality adjectives. These are more frequently used in both American (14.43%) and British (13.99%) tourism texts than in Greek ones (3.12%), but target texts appear to stay very close to their source texts (4.50%). The aspect of Quality, and its implied element of competition, as a promotional tool has been found to play a key role in English tourism discourse, something that, however, does not seem to have been taken into account when translating from Greek. This lack of adaptation is also manifested in the most frequently used adjective in this category, which is *ideal*, a translation of *ιδανικός*, also the most frequently used adjective in Greek tourism texts. In contrast, the most frequently used Quality adjective in English is *good*, while *ideal* has not been found in either American or the British tourism texts. Moreover, the frequency with which *ideal* and *ιδανικός* are being used is almost identical (82 instances in the source texts and 80 in the target texts). There is strong and clear evidence that target texts stay very close to their source texts regarding the semantic category of Quality.

Finally, a note is needed regarding Beauty adjectives, since their case is less straightforward as a result of the difference in the distribution of this semantic category between American and the British tourism texts. This semantic category of adjectives is used with a similar frequency in both Greek tourism texts (11.86%) and American ones (9.01%), but with a much higher one in British English texts (18.16%). Their frequency slightly drops in the target texts (9.34%) compared to the source texts, which suggests adaptation is in the opposite direction compared to British English, but not required and not observed compared to American English. Because of this rare situation, also noted in the case of the French-English translations, it is not possible to argue that target texts follow the patterns found in American English tourism texts, rather than British English ones, which would explain the frequencies above. Instead, considering the generally similar frequencies between the source and the target texts and the fact that no adaptation is required when Greek tourism texts are compared to American English texts regarding this semantic category, it is concluded here that adaptation is not required and indeed not clearly observed. Nevertheless, this does not mean that the target texts stay very close to their source texts, and instances of omission are also present, which would explain the slight drop in the frequency. As illustrated in Example 10.6, an entire sentence which includes a number of descriptions of physical space, including three through the lens of Beauty, has been completely omitted. The reasons behind this omission are unclear, it might have been considered a rather 'loaded' sentence with too much emphasis on Beauty.

Example 10.6

ST *Είναι ιδιαίτερα **γραφική**, με καθαρές θάλασσες, φημισμένα μοναστήρια, σπουδαίους αρχαιολογικούς χώρους, **γοητευτική** πρωτεύουσα και **όμορφα χωριά.***

BT *It is particularly **picturesque**, with clear seas, famous monasteries, great archaeological sites, a **charming** capital and **beautiful villages.***

TT –

Source www.visitgreece.gr

Overall, no attempt towards adaptation is noted in the case of English translations from Greek texts, and perhaps more importantly no clear pattern can be identified at least in the case of Size adjectives, something that has also been observed during the previous case study for this particular language pair. The resulting tourism discourse does not therefore manage to conceptualise and communicate aspects of physical space in line with how English non-translated tourism texts conceptualise these.

10.1.3 Translations into English from Russian

Even though a very high degree of adaptation might be expected in English translations from Russian, as a result of the considerable differences between Russian and English non-translated tourism texts, adaptation is actually quite minimal, with most semantic categories (i.e. Size, Time, Beauty, and Quality) not being adapted. Successful adaptation is noted only for Fame adjectives. Figure 10.3 summarises the quantitative findings presented earlier in Table 10.2.

As noted, successful adaptation is only observed in the case of Fame adjectives. These are less frequently used in Russian tourism texts (5.07%) than in both American (13.30%) and British (10.24%) texts, while their frequency has approximately doubled during translation into English (9.80%), resulting in no significant difference when English (both varieties) non-translated tourism texts are compared with English translations from Russian. There appears to be a clear recognition that Fame adjectives play an important role in English tourism discourse and that they should therefore be used more frequently during translation. However, even with this recognition in mind, the adjectives used are only

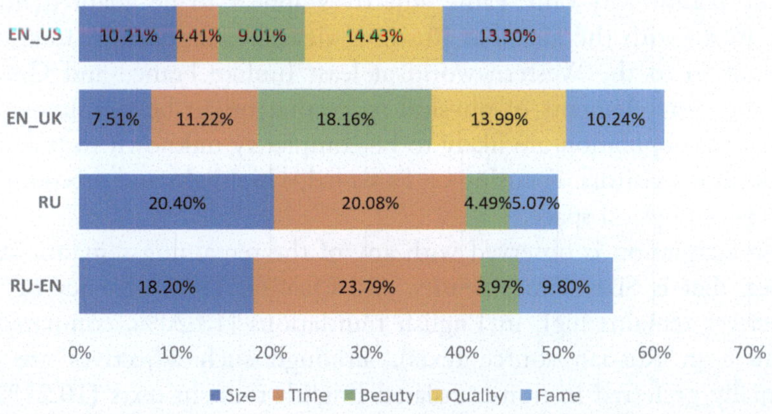

Fig. 10.3 Normalised frequencies of most prominent semantic categories of descriptive adjectives in non-translated English, Russian source texts, and translated English from Russian as % of all identified adjectives

those considered prototypical of this category, that is *famous* and *popular*, which are equivalents of the Russian *известный* and *популярный*, even though *iconic* is also used in English tourism discourse. Based on this, it can be argued that there is a reluctance to increase the subjectivity potentially associated with this semantic category. Example 10.7 illustrates how a Fame adjective has been added during a slight reorganisation of the sentence, shifting attention away from the Size of the bridge to what the river is famous for. Here, not only has there been shift in terms of the semantic category, but also the attention of the reader is directed towards a different element of physical space.

Example 10.7

ST *У Исаакиевского сквера через реку Мойку перекинут Синий мост – один из "цветных" мостов и самый широкий мост в Петербурге.*
BT *At St. Isaac's Square, the Blue Bridge spans over the Moika River - one of the "colored" bridges and the widest bridge in St. Petersburg.*
TT *The composition of the square includes the Blue bridge connecting the two banks of the* ***Moika river famous*** *for its series of "bridges of color".*
Source www.visit-petersburg.ru

The reason why only Fame adjectives appear to be adapted might have to do with the fact that Russia is viewed as a generally unknown destination to the Western world at least (unlike France and Greece) and therefore elements of physical space that might be well-known to the native population are likely to be completely unknown to potential Anglophone visitors. Focusing on Fame helps highlight and promote key aspects of physical space.

No adaptation is observed with any of the remaining semantic categories, that is Size, Time, Beauty, and Quality. The frequency of Size adjectives remains high in English translations (18.20%, compared to 20.40% in Russian source texts), although such adjectives are not normally preferred in non-translated English tourism texts (10.21% in American English and 7.51% in British English). However, some small recognition that adaptation might be required here is evidenced by the fact that only the two most prototypical Size adjectives are used in translated texts, namely *large* and *small*, even though other options such as *просторный* (*spacious*) and *высокий* (*high*) are also found in Russian source texts. Focusing on more typical considerations of size at least provides the English-speaking reader-tourists with information that they might also prioritise, albeit not to the same degree.

A very similar picture can be observed with Time adjectives. These are used with approximately the same frequency in both the Russian source texts (20.08%) and their English target texts (23.79%) which is much higher than what is found in both American English (4.41%) and British English (11.22%) non-translated tourism texts. This is further manifested by equivalent adjectives being used in both the source and target texts, most notably the use of *modern* which has not been found in English non-translated tourism texts. There appears to be no recognition of the need to reduce the prominence of objective descriptions (and conversely increase that of subjective ones) in English tourism discourse.

Finally, adaptation is also not noted in the case of Quality adjectives, with their frequency being 0% in both the source and the target texts compared to the much higher frequencies found in both American (14.43%) and British English (13.99%) tourism texts. This provides clear evidence that translators are reluctant to move away from the source text which in the case of Quality adjectives would mean adding such descriptions of physical space. Very similar observations are made when examining the other subjective category, that is Beauty. Frequencies in the source (4.49%) and the target texts (3.97%) are close to each other, and much lower than what is found in American English (9.01%) and British English (18.16%) non-translated tourism texts. Only the most prototypical beauty adjective is used in the translated texts, that is *beautiful*, very similar to what has been found in Russian non-translated texts. In contrast, three different Beauty adjectives are used in American English tourism texts, and 7 in British ones. The importance of subjective descriptions for the promotional effect of tourism discourse in English appears to have not been recognised.

In sum, a very limited degree of adaptation is observed in English translations from Russian and there is almost no effort to make the target text represent and communicate a message about physical space in accordance with how an Anglophone audience might gaze at the world, even though differences between English and the Russian non-translated tourism texts are particularly pronounced and therefore the need for adaptation could have been perhaps more easily identified. The resulting text is an English version of how Russian speakers categorise physical space, rather than a description of the destination through the eyes of an English speaker.

Overall, in the case of translations of tourism texts into English, although generally high degrees of adaptation might be expected in most language pairs (with the potential exception of Greek-English), there is generally a reluctance to move away from the source text. Some attempts towards adaptation, albeit mainly weak, are observed in the case of French-English mostly with subjective semantic categories. English translations of Greek texts present a rather unclear pattern with evidence of reverse adaptation, while translations from Russian only successfully adapt Fame adjectives. Therefore, although specific parts of the English target texts might attempt to recreate an image of physical space as conceptualised by English-speaking reader-tourists, overall there appears to be little recognition of the importance of adaptation.

10.2 Tourism Texts Translated into French

Having established the adaptation patterns in the tourism texts translated into English, the analysis now focuses on texts translated into French. Compared to English as a target language, a smaller number of websites are expected to have a French version, even though the TrAIL corpus has been designed in such a way to maximise multilingual data. Specifically, although all English websites (both American and British) examined here offer a French version, only two Greek ones do, while no translations into French from Russian official tourism websites have been identified. While this affects to some extent the comparability of results, it is considered an inevitable practical limitation of a multilingual corpus design. Results are in all cases normalised and therefore comparisons can be made even across corpus components that are uneven in terms of size.

As before, the first thing to examine is the number of adjectives, both descriptors and classifiers, employed in tourism texts which meet the threshold frequency measure of 0.18ptw and which also modify nouns denoting physical space in non-translated texts and translated texts and their source texts, to help establish the basis for the calculations of the frequency distribution of semantic categories (Table 10.4).

As can be seen, the distribution of identified adjectives for the 'total' number, is similar between the source and the target texts, with the exception of Greek into French translations. This might be perceived as an indication that not much adaptation might be taking place when tourism texts are translated into French.

Next, the distribution of the most prominent semantic categories of descriptive adjectives out of all 'top' adjectives is examined (Table 10.5).

The next table (Table 10.6) summarises the adaptation patterns identified in the translations into French.

As with translations into English, a mixed picture emerges regarding translations into French. Even though adaptation is generally required, there is only one case of successful adaptation, and a few cases of weak adaptation. Overall, target texts appear to be staying very close to their respective source texts, as was generally the case with the translations into English. To get a better understanding of how each language pair has adapted representations of physical space, a closer look at how each individual pair has dealt with each semantic category is required.

Table 10.4 Normalised frequencies of most frequent adjectives modifying nouns denoting physical space in non-translated French, source texts (i.e. non-translated texts in American English, British English, Greek), and translated French corpus components

		Tokens	Adjectives (types)	Adjectives (tokens)	
				f	nf (ptw)
Non-translated texts	FR	170,060	29	1,791	10.532
Source texts	EN_US	170,520	27	1,587	9.30
	EN_UK	170,704	47	2,637	15.44
	EL	170,601	57	3,812	22.34
Translated texts	*EN_US-FR*	196,771	21	1,592	8.09
	EN_UK-FR	128,742	39	1,996	15.50
	EL-FR	130,623	34	1,910	14.62

Table 10.5 Normalised frequencies of most prominent semantic categories of descriptive adjectives in non-translated French, source texts (i.e. non-translated texts in American English, British English, Greek), and translated French corpus components as a percentage of identified verbs (tokens)

		NT TEXTS (TL)	NT TEXTS (SL)			TRANSLATED TEXTS		
		FR	EN_US	EN_UK	EL	EN_US-FR	EN_UK-FR	EL-FR
Total number of adjectives		1,791	1,587	2,637	3,812	1,592	1,996	1,910
Size	Raw frequency	403	162	198	583	251	234	416
	Normalised frequency	22.50%	10.21%	7.51%	15.29%	15.77%	11.72%	21.78%
Time	Raw frequency	269	70	296	519	96	167	299
	Normalised frequency	15.02%	4.41%	11.22%	13.61%	6.03%	8.37%	15.65%
Beauty	Raw frequency	149	143	479	452	150	361	213
	Normalised frequency	8.32%	9.01%	18.16%	11.86%	9.42%	18.09%	11.15%
Quality	Raw frequency	0	229	369	119	84	120	78
	Normalised frequency	0%	14.43%	13.99%	3.12%	5.28%	6.01%	4.08%
Fame	Raw frequency	87	211	270	384	224	156	255
	Normalised frequency	4.86%	13.30%	10.24%	10.07%	14.07%	7.82%	13.35%

Table 10.6 Patterns of adaptation in translations into French (descriptive adjectives)

	EN_US-FR	EN_UK-FR	EL-FR
Not Required	Beauty		Time Beauty
Additional			
No	Time Fame	Time Beauty Fame	Quality
Successful			Size
Weak	Size Quality	Size Quality	
Over			
Reverse			Fame

10.2.1 Translations into French from American English

For translations from American English into French, adaptation might be expected for all semantic categories, apart from Beauty, where it is indeed not observed. No adaptation is noted for Time and Fame adjectives, while some attempt towards adaptation, albeit weak, is observed in the semantic categories of Size and Quality. Figure 10.4 summarises the quantitative findings presented earlier in Table 10.6.

Some attempt towards adaptation is observed in the category of Size adjectives, albeit weak. Adjectives belonging to this category are found twice as frequently in French non-translated tourism texts (22.50%) than in American English ones (10.21%). While their frequency increases during translation into French (15.77%), it does not reach the levels found in non-translated texts, based on the inferential statistical tests. There seems to be, however, some recognition that objective descriptions based on Size are particularly important in French tourism discourse—it is indeed the most prominent semantic category. Example 10.8 illustrates how adaptation of Size adjectives is realised. Here, the entire paragraph about *North Valleyheart Riverwalk* has been rewritten during translation with the focus shifting away from what one can do there to a description of the destination. Once again, it appears that there is more room for adaptation when translators have the freedom to break away from

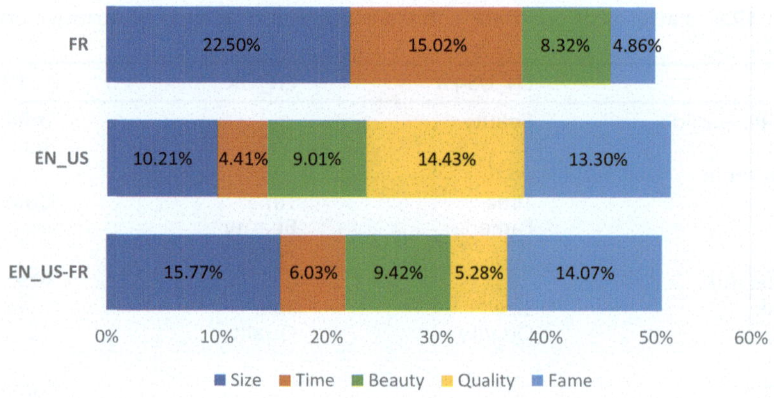

Fig. 10.4 Normalised frequencies of most prominent semantic categories of descriptive adjectives in non-translated French, American English source texts, and translated French from American English as % of all identified adjectives

the source text, which is evidenced by the considerable shifts during translation.

Example 10.8

ST *North Valleyheart Riverwalk is a 1.9-mile linear park along a section of the flood control channel that offers healthy exercise in relaxed surrounds plus a few shaded spots for pause or a picnic.*

TT *Faites l'expérience de la revitalisation du Los Angeles River en effectuant une balade le long de l'étendue de 1,9 miles à Studio City, encadrée par deux* ***petits parcs.***

BT *Experience the revitalization of the Los Angeles River with a stroll along the 1.9-mile stretch in Studio City, framed by two* ***small parks.***

Source www.discoverlosangeles.com

Regarding the use of *petit* (*small*), it must be noted that it is used with approximately half the frequency in French translations from American English compared to French non-translated texts, which reveals that there might not be a recognition of the key role that this particular adjective plays in French. Similarly, other adjectives such as *haut* (*tall*) and *vaste* (*vast*) are not found in the target texts—the emphasis appears to be only on the aspect of dimension, i.e. big/small, similar to what has been noted in American English non-translated tourism texts. All this provides evidence that adaptation is indeed weak.

Another semantic category where weak adaptation is noted is Quality. This category has not been found in French non-translated texts, while it is fairly frequently used in American English texts (14.43%). A significant decrease is noted when these texts are translated into French (5.28%) which suggests some recognition that this category does not play a key role in how physical space is conceptualised in French tourism discourse. Interestingly, the only item used to realise this category in target texts is *meilleur* (*best*), which clearly highlights the element of competitiveness associated with Quality adjectives (Example 10.9). This focus might affect how the target text is received by the French-speaking reader-tourists, who might not be accustomed to categorising physical space based on such competitive elements.

Example 10.9

ST *Some of the biggest and* ***best malls*** *in the USA also call the area home.*

TT *C'est aussi là que sont installés quelques-uns des plus grands et* ***des meilleurs centres commerciaux*** *du pays.*

BT *It's also home to some of the country's biggest and* ***best shopping malls.***

Source www.visittheusa.com

Finally, no adaptation is noted for Time and Fame adjectives, with their frequencies staying very close to those found in their American English source texts. Regarding Time adjectives, these are used less frequently in American English tourism texts (4.41%) than in French ones (15.02%), while their frequency remains low during translation (6.03%). Interestingly, the only Time adjective employed in the target texts is *ancien* (*ancient*), which would be the equivalent of the only Time adjective found in American English tourism texts, and not *vieil* (*old*), which is however found in non-translated French tourism texts. The reverse pattern is observed regarding Fame adjectives, which are found more frequently in American English tourism texts (13.30%) than French ones (4.86%), and a very similar frequency has been found in target texts (14.07%) suggesting a close relationship between the source and the target texts. This is further supported by the fact that exactly the 'same' adjective types are found in both the source and the target texts, indicating a very close translation.

Overall, there is little evidence of adaptation, which when noted is weak, suggesting that largely translators are reluctant to move away from American English source text when translating into French. This can have a significant impact on how a destination is communicated to the French-speaking reader-tourist and the extent to which they might find descriptions appealing and in line with how they themselves gaze at and categorise the world around them.

10.2.2 Translations into French from British English

Turning now to the translations from British English, we can see that similar to the translations from American English, although a generally high degree of adaptation might be expected, there is very little evidence of adaptation, with only Size and Quality adjectives demonstrating weak adaptation. Overall, it seems that, as with the other variety of English, the British English translations into French stay relatively close to their source texts. Figure 10.5 summarises the quantitative findings presented earlier in Table 10.6.

A weak attempt towards adaptation is observed with Size adjectives. These are used significantly more frequently in French tourism texts (22.50%) than in British English ones (7.51%) reflecting the important role that objective descriptions play in French tourism discourse. While an increase in their frequency is noted during translation (11.72%) this is still approximately half what it is in French non-translated texts, suggesting a rather big gap between the two types of text. While this weak adaptation affects all individual adjectives used, it is perhaps most evident in the case of *petit* (*small*), which, as has been explained in the previous chapter, plays a key role in French tourism discourse by having a particularly positive semantic prosody not shared by its equivalent in English. One might therefore expect to see an increase in its use in target texts, which is indeed observed: their frequency rises from 27% of all

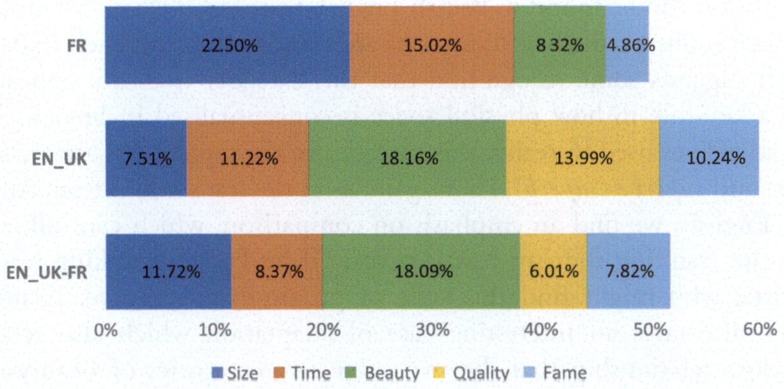

Fig. 10.5 Normalised frequencies of most prominent semantic categories of descriptive adjectives in non-translated French, British English source texts, and translated French from British English as % of all identified adjectives

Size adjectives in British English source texts to 35% in French target texts (compared to 38% in French non-translated texts). Example 10.10 illustrates how this particular adjective is often added in target texts. It is perhaps no surprise that such additions are found in cases where the translators have the freedom to move away from the source text, as evidenced by the number of shifts in the target text.

Example 10.10

ST *The tiny, enchanting Isle of Iona is a peaceful haven, while you can discover one of Scotland's geological wonders on a boat trip to Staffa…*

TT *La minuscule et charmante île de Iona est un paradis paisible, avec ses jolies* ***petites baies*** *et son ancienne abbaye. Admirez lune des merveilles géologiques de l'Écosse lors d'une visite en bateau de Staffa…*

BT *The tiny, charming island of Iona is a peaceful paradise, with its pretty* ***little bays*** *and ancient abbey. Admire one of Scotland's geological wonders on a boat tour of Staffa…*

Source www.visitscotland.com

The other semantic category where weak adaptation is observed is Quality. This category has not been found in French non-translated texts, but it can be found in British English texts (13.99%). A significant decrease is observed when these texts are translated into French (6.01%) which suggests some recognition that this category does not appear to play a key role in how physical space is conceptualised in French. The two adjectives used to realise this category in the target texts are *meilleur* (*best*) and *superbe* (*superb*). Here again, as in the translations from American English, we find an emphasis on comparison, which can influence how the resulting tourism text is perceived by French-speaking reader-tourists, who might find this style of promotion aggressive. Example 10.11 illustrates an interesting case of adaptation which also reflects the close relationship that the two subjective categories of Beauty and Quality have and how they might be (re)interpreted by the translator(s). Here, the Beauty adjective *magnificent*, which could have been literally translated as *magnifique*, has instead been rendered as the Quality adjective *superbe* highlighting the grandeur of the building especially in comparison with others. While generally Quality descriptions are omitted in the target texts, this is a notable example of such a description being added, which suggests that the distinction between Beauty and Quality, and the role these semantic categories play in tourism discourse, is not always clear-cut and much more open to individual interpretation compared to more objective descriptions.

Example 10.11

ST *This **magnificent building** contains an evocative Great Hall, fine furniture and paintings.*

TT *Ce **superbe bâtiment** contient une grande salle, des peintures et un mobilier élégants.*

BT *This **superb building** features a large hall, elegant paintings and furnishings.*

Source www.visitscotland.com

Finally, no adaptation is noted for the semantic categories of Time, Beauty, and Fame, even though it might be expected. It is interesting that there is a mix of both objective and subjective categories here, indicating that the target texts stay close to their source texts regardless of the semantic category. Specifically, Time adjectives are used less frequently in both the source (11.22%) and the target texts (8.37%) than in French non-translated texts (15.02%). Some small shifts are, however, noted, which might be evidence of the recognition that some adjustments are required during translation. For instance, although *new* is found in British English source texts, its equivalent (*nouveau*) is not found in French target texts, and it was also not found in French non-translated tourism texts. The reverse pattern is observed for Beauty and Fame adjectives. Beauty adjectives are much more frequently found in British English (18.16%) than French (8.32%) tourism texts, something that is not reflected in the French target texts (18.09%). Furthermore, a similar number of adjective types is used in source and target texts (7 vs 5), many more than is found in French non-translated texts (only 2), which is a reflection of the emphasis put on Beauty in translated texts from British English. Similarly, Fame adjectives are scarcely used in French tourism discourse (4.86%), but they are almost three times more frequently found in French target texts (14.07%), a frequency closer to that found in their British source texts (10.24%). In terms of particular adjectives, close equivalents of the English ones are found in French target texts, including *emblématique* (*iconic*)—a very typical Fame adjective of English tourism discourse (both American and British).

Overall, the main adaptation pattern observed in British English-French translation is that of a reluctance to move away from the source texts. Some limited evidence of adaptation is available, but adaptation is weak at best. It is worth noting that where some adaptation is noted it is for the categories that are most (i.e. Size) and least (i.e. Quality) prominent in French non-translated tourism texts respectively. This could be seen as an indication that where extremes are present, it is easier to identify some need for adaptation or that there is more scope for adaptation to occur where differences are more pronounced, although data to confirm this are limited. In any case, the resulting target text is still quite different from how French-speaking reader-tourists conceptualise their world around them and is likely to have an impact on how it is received by them.

10.2.3 Translations into French from Greek

A generally medium degree of adaptation might be expected with Greek-French translations, as a result of the similarities between French and Greek non-translated tourism texts established during the previous stage of analysis in the semantic categories of Time and Beauty, where indeed no adaptation is observed. Successful adaptation is noted only in the case of Size adjectives, while Quality adjectives are not adapted. Finally, Fame adjectives appear to be adapted in the opposite direction. Overall, no clear pattern for adaptation is noted. Figure 10.6 summarises the quantitative findings, presented earlier in Table 10.6.

As mentioned above, successful adaptation is noted only in the case of Size adjectives. Although prominent in both French and Greek tourism texts, they are more frequently found in French texts (22.50%) than in Greek ones (15.29%). Their frequency increases in translated texts (21.78%) and comes very close to that found in French non-translated texts. To examine how exactly this adaptation might be realised, it is useful to look once again into *petit* (*small*) which as mentioned before has a particularly positive semantic prosody in French.

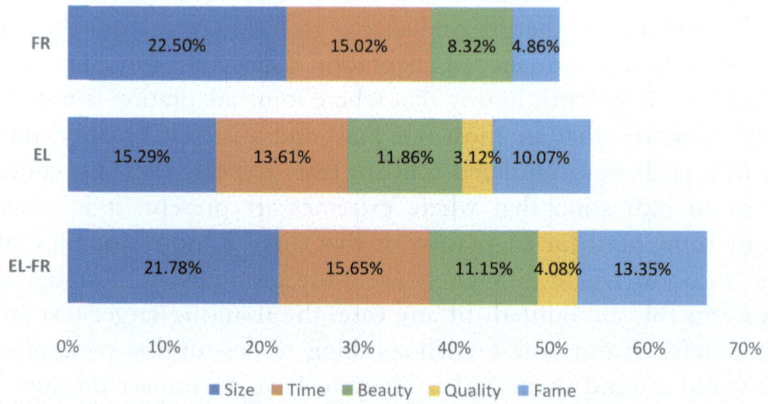

Fig. 10.6 Normalised frequencies of most prominent semantic categories of descriptive adjectives in non-translated French, translated French from Greek, and Greek source texts as % of all identified adjectives

Example 10.12 showcases how this adjective replaces the Time adjective παραδοσιακός (*traditional*) as a recognition of the key role that it plays in French tourism discourse.

Example 10.12

ST *Ό,τι έχετε φανταστεί για τα νησιά του Αιγαίου των δεκαετιών του'60 και του'70 είναι ζωντανό εδώ: αυθεντικά χωριά,* ***παραδοσιακά καφενεία****, εκκλησάκια θεμελιωμένα στους βράχους, πλατείες για ουζάκι και μεζέ, ξερολιθιές, θερμά λουτρά και μονοπάτια.*

BT *Everything you have imagined about the Aegean islands of the 60s and 70s is alive here: authentic villages,* ***traditional cafes****, small churches founded on the rocks, squares for ouzo and snacks, dry stone walls, thermal baths and paths.*

TT *Tout ce dont vous avez pu imaginer des îles de la mer Égée des années'60 et'70 est encore vivant ici: les villages traditionnels, les* ***petits cafés****, les petites églises accrochées aux rochers, les places centrales où l'on déguste de l'ouzo avec des mézés (hors-d'œuvre), des murets de pierres, des eaux thermales et des sentiers étroits.*

BT *Everything you might have imagined about the Aegean islands of the '60s and '70s is still alive here: traditional villages,* ***little cafés****, little churches clinging to the rocks, central squares where you can enjoy ouzo with mezes (hors d'oeuvres), low stone walls, thermal waters and narrow paths.*

Source www.discovergreece.com

A rather unique case is that of Fame adjectives, which appear to have been adapted in the opposite direction. Fame adjectives are employed almost twice as frequently in Greek tourism texts (10.07%) compared to French ones (4.86%), but their frequency instead of decreasing during translation, actually slightly increases (13.35%). However, the frequency is still fairly close to that found in Greek source texts and close inspection of the concordance lines reveals that indeed target texts stay very close to their source texts when it comes to Fame adjectives. Further evidence is obtained by the presence of *important* in the target texts as a translation of σημαντικός (*significant*) and σπουδαίος (*important*). This emphasis on importance is a unique feature of Greek tourism discourse which has been transferred in the translations.

No instances of adaptation, even though they might be expected, are found with Quality adjectives. Such adjectives have not been found in French non-translated tourism texts but they have been found to be used, albeit scarcely, in both the Greek source texts (3.12%) and their French target texts (4.08%). This is further manifested by the fact that equivalent adjectives are used in both the source and the target texts, i.e. καλός (*good*) and ιδανικός (*ideal*) and *bon* (*good*) and *idéale* (*ideal*). It is worth noting that *idéale* (*ideal*) is not found in French non-translated texts. Although the frequencies are generally low, there is evidence to suggest that the resulting target text does not correspond to the communicative conventions of French tourism discourse and to how French-speaking reader-tourists interpret the world around them.

Overall, little adaptation is noted in the case of French translations from Greek, with target texts staying generally close to their source texts in the case of Quality adjectives, while a lack of a clear pattern is noted with Fame adjectives. The only notable exception to this and where successful adaptation is observed in the case of Size adjectives, which could be explained by the fact that it has been recognised as a category that is prominent in French tourism discourse particularly through the distinctive use of *petit*. Nevertheless, the resulting text still largely reflects the way in which Greek reader-tourists might perceive the world around them which can have significant implications for the promotional potential of the target texts.

In the case of the translations of tourism texts into French, although generally high degrees of adaptation might be expected in most cases, there are few attempts to move away from the source text, with the vast majority of them considered weak. Such is the case with French translations from English sources, with both varieties exhibiting almost identical patterns. The only instance of successful adaptation is observed with French translations of Greek texts, but the target texts in this language pair also adapt adjectives in the opposite direction, and therefore, the pattern of adaptation is not clear. The translations from Greek sources, irrespective of the target language, appear to be the only ones where this lack of a clear pattern has been observed so far. As with translations into English, even though some parts of the texts might appear to have been adapted, overall the resulting target texts represent an image of physical space that is still far away from how French-speaking reader-tourists might gaze at the world around them.

10.3 Tourism Texts Translated into Russian

The last part of this case study focuses on the examination of the adaptation patterns in the tourism texts translated into Russian. It is important to note that the corpus components involving translations into Russian are rather small due to the availability of data and, although results are normalised, caution is required when interpreting the results. As in previous sections, the starting point is to establish the total number of adjectives (both descriptors and classifiers) describing physical space to identify the basis for the calculations of the normalised frequencies of the most prominent semantic categories of descriptive adjectives, that is, the number of adjectives employed in tourism texts which meet the threshold frequency measure of 0.18ptw and which also modify nouns denoting physical space in non-translated texts and translated texts and their source texts (Table 10.7).

Although not the focus of this case study, it is interesting to note that, apart from the language pair Greek-Russian, many more adjectives have been found in the translated texts compared to both source texts

Table 10.7 Normalised frequencies of most frequent adjectives modifying nouns denoting physical space in non-translated Russian, source texts (i.e. non-translated texts in American English, British English, French, Greek), and translated Russian corpus components

Tokens			Adjectives (types)	Adjectives (tokens) f	nf (ptw)
Non-translated texts	RU	170,881	34	1,892	11.07
Source texts	EN_US	170,520	27	1,587	9.30
	EN_UK	170,704	47	2,637	15.44
	FR	170,060	29	1,791	10.53
	EL	170,601	57	3,812	22.34
Translated texts	EN_US RU	53,430	42	846	15.83
	EN_UK RU	43,707	62	1,040	23.79
	FR-RU	36,464	49	790	21.66
	EL-RU	88,461	62	2,280	25.77

and the Russian non-translated tourism texts, irrespective of the source language. This might reflect the small corpus component size, with the possible exception of the Greek-Russian component. The smaller the corpus component, the larger the minimum number of occurrences for each adjective for it to be included in the analysis. However, the fact that a similar 'pattern' has not been observed when examining verbs in the previous case study suggests that this is unlikely to be a methodological limitation and that indeed the translated texts tend on the whole to use more adjectives.

The distribution of the most prominent semantic categories of descriptive adjectives out of all 'top' adjectives is summarised in Table 10.8.

The next table (Table 10.9) summarises the adaptation patterns identified in the translations into Russian.

Adaptation might be expected in most cases across all language pairs, apart from French-Russian, reflecting the fact that Russian tourism discourse generally stands out from such discourse in English and Greek, but much less so from French. While this creates more opportunities for adaptation, only one, weak, attempt towards it is noted, unlike what has been observed during the previous case study of verbs. It is rather clear in the case of the translations into Russian that the target texts stay extremely close to their source texts.

10.3.1 Translations into Russian from American English

Although adaptation is required for all semantic categories when American English tourism texts are translated into Russian, due to the differences between non-translated American English and Russian tourism texts, as evidenced in Chapter 9, no instances of adaptation are noted. The resulting target text communicates an image of physical space that reflects the American society rather than the Russian one. Figure 10.7 summarises quantitative findings presented earlier in Table 10.8.

As noted, no adaptation is observed in any of the semantic categories of Size, Time, Beauty, Quality, and Fame. Size adjectives in Russian

Table 10.8 Normalised frequencies of most prominent semantic categories of descriptive adjectives in non-translated Russian, source texts (i.e. non-translated texts in American English, British English, French, Greek), and translated Russian corpus components as a percentage of identified verbs (tokens)

		RU	EN_US	EN_UK	FR	EL	EN_US-RU	EN_UK-RU	FR-RU	EL-RU
Total number of adjectives		1,892	1,587	2,637	1,791	3,812	846	1,040	790	2,280
Size	Raw frequency	386	162	198	403	583	60	105	134	361
	Normalised frequency	20.40%	10.21%	7.51%	22.50%	15.29%	7.09%	10.10%	16.96%	15.83%
Time	Raw frequency	380	70	296	269	519	31	99	98	273
	Normalised frequency	20.08%	4.41%	11.22%	15.02%	13.61%	3.66%	9.52%	12.41%	11.97%
Beauty	Raw frequency	85	143	479	149	452	75	195	90	278
	Normalised frequency	4.49%	9.01%	18.16%	8.32%	11.86%	8.87%	18.75%	11.39%	12.19%
Quality	Raw frequency	0	229	369	0	119	105	41	15	72
	Normalised frequency	0%	14.43%	13.99%	0%	3.12%	12.41%	3.94%	1.90%	3.16%
Fame	Raw frequency	96	211	270	87	384	96	89	64	271
	Normalised frequency	5.07%	13.30%	10.24%	4.86%	10.07%	11.35%	8.56%	8.10%	11.89%

Table 10.9 Patterns of adaptation in Russian translations (descriptive adjectives)

	EN_US-RU	EN_UK-RU	FR-RU	EL-RU
Not Required			Size Fame	
Additional			Quality	
No	Size Time Beauty Quality Fame	Size Time Beauty Fame	Time Beauty	Size Time Beauty Quality Fame
Successful				
Weak		Quality		
Over				
Reverse				

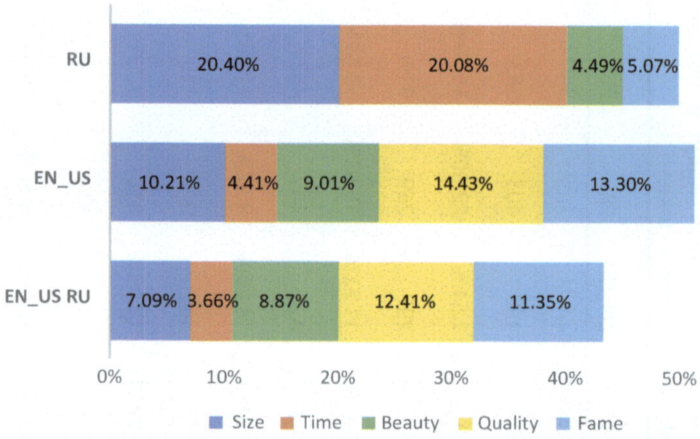

Fig. 10.7 Normalised frequencies of most prominent semantic categories of descriptive adjectives in non-translated Russian, translated Russian from American English, and American English source texts as % of all identified adjectives

translations (7.09%) are not adapted to match the higher frequency with which they are used in Russian (20.40%) non-translated tourism texts, and they remain closer to the frequencies found in American English source texts (10.21%). It could be possible to argue that some evidence of adaptation might, however, be noted here, as a result of the use of *обширный* (*vast*) found in the target texts. The only Size adjectives found in American English source texts are those related to dimensions (i.e. *big/small*), not extent, while Russian tourism texts focus on other aspects of physical space as a result of their much stronger focus on this semantic category. However, in all 10 instances where *обширный* (*vast*) has been found, an equivalent adjective is found in the source texts. This is not considered an instance of adaptation, but rather a reflection and limitation of the small corpus component size, as the smaller it is the lower the minimum number of adjectives that needs to be captured.

A very similar pattern is noted in the case of Time adjectives. Although they represent a high frequency in Russian non-translated tourism texts (20.08%), the frequency found in the target texts (3.66%) reflects that found in the source texts (4.41%). As with Size adjectives, some evidence of adaptation might be found behind the use of the adjective *модный* (*modern*) in the target texts, when the only Time adjective found in the source texts is *old*. However, as with Size, this is again a reflection and limitation of the small corpus component size rather than a genuine instance of adaptation. As far as the two objective semantic categories of adjectives are concerned, Size and Time, which are highly emphasised in Russian tourism discourse, the translated texts have not succeeded in highlighting them. The resulting tourism texts does not reflect a representation of the world in-line with how Russian-speaking reader-tourists perceive it, but rather encode a very American way of conceptualising and classifying physical space.

A similar lack of adaptation is noted even with the more subjective semantic categories of adjectives, such as that of Beauty. Contrary to the objective categories, Beauty adjectives are scarcely used in Russian non-translated tourism texts (4.49%), while they are more frequently found in Russian translations (8.87%) with a very similar frequency to that noted in the source texts (9.01%). An even more obvious lack of adaptation is also noted with the other subjective category, i.e. Quality adjectives. Such adjectives have not been found in Russian tourism texts, but they are found in American English texts (14.43%). Target texts employ these with a similar frequency to that found in source texts (11.72%). Emphasis is therefore placed on how good aspects of the destination are, often with a competitive element associated with this category, which is something that is not particularly prominent in Russian tourism discourse. Finally, no adaptation is noted for Fame adjectives, which are used with approximately the same frequency in both source (13.30%) and target texts (11.35%), although they are more rarely found in Russian non-translated tourism texts (5.07%).

Overall, even though a high degree of adaptation is required, no attempt is observed towards it in the case of Russian translations from American English, with all semantic categories staying close to the source texts. However, it must be noted that translations from only one American tourism website are included in this Russian component, and therefore any generalisations should be made with caution. The limited evidence obtained here suggests that the resulting translated tourism texts represent a rather American view of the world, which is very distant from how physical space is conceptualised in the Russian society. If this is to be considered representative of this language pair more generally, then it could either be interpreted as a lack of appreciation of the need for adaptation for a new audience or a reflection of the prestigious status of the American culture.

10.3.2 Translations into Russian from British English

Russian translations from British English present a very similar picture to that observed in the analysis of American English-Russian. Although adaptation might be expected for all semantic categories, based on the findings in Chapter 9, it is only observed in the case of Quality adjectives, and even then it is considered to be weak. Overall, the target texts are generally reluctant to move away from the source texts. Figure 10.8 summarises the quantitative findings presented earlier in Table 10.8.

Some evidence of successful adaptation is noted in the case of Quality adjectives, albeit weak. These adjectives have not been found in Russian non-translated tourism texts, but they are the second most prominent category in British English texts (13.99%). During translation, their frequency significantly drops (to 3.94%) but they are still present. This is realised mainly by omission or replacement, as can be seen in Example 10.13. In this example, two Quality adjectives are used in the source text to describe Belfast. The first of these, *superb*, is replaced during translation by an adjective belonging to the Other category, i.e. *роскошный*

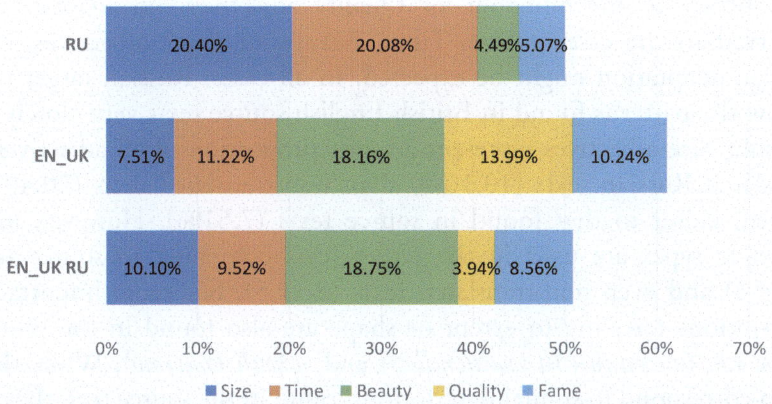

Fig. 10.8 Normalised frequencies of most prominent semantic categories of descriptive adjectives in non-translated Russian, British English source texts, and translated Russian from British English as % of all identified adjectives

(*luxurious*). The second adjective, *top*, together with the reference to attractions has been omitted in the target text. Although adaptation is considered weak for this category of adjectives, the decrease in their frequency is considerable and it is only because of the 0% found in the Russian non-translated texts that it cannot be considered successful. It can be assumed that for semantic categories where most extreme differences are noted, it is more likely that the need for adaptation might be recognised.

Example 10.13

ST *A visit will swiftly be rewarded with welcoming locals,* ***superb pubs and restaurants****, and* ***top attractions*** *including the atmospheric Crumlin Road Gaol and walking tours of Belfast's famous murals.*

TT *Будьте готовы к встрече с приветливыми местными жителями,* **роскошными пабам и ресторанам,** *а также к необычным развлечениям.*

BT *Be prepared for friendly locals, luxurious pubs and restaurants, and unusual entertainment.*

Source www.visitbritain.com

Contrary to the outcome for Quality adjectives, no adaptation is observed in the case of Size, Time, Beauty, or Fame adjectives, even though adaptation might be expected. In all cases, Russian target texts follow the patterns found in British English source texts very closely. As a result, Size adjectives represent a lower proportion of all adjectives in translated Russian texts (10.10%) than non-translated texts (20.40%), indeed, closer to that found in source texts (7.51%). However, more adjective types are used in the target texts compared to source texts (8 vs 4) and even non-translated texts (8 vs 5), but more importantly descriptions referring to extent or shape are also found in the translations, i.e. *бескрайний* (*vast/endless*) and *узкий* (*narrow*). While these often correspond to similar descriptions found in the source text, they are also added to enhance the target text, as in Example 10.14, where additional information includes a Size description, once again confirming that when translators have more freedom to move away from the source text, as evidenced from the number of shifts, there are more chances for adaptation to occur. However, such instances of 'rewriting' are rare and therefore the total frequency of Size adjectives is not affected in any meaningful way.

Like Size, Time adjectives are more frequently used in Russian non-translated texts (20.08%) than British English ones (11.22%), but their frequency in the target texts (9.52%) is closer to that found in the source texts. As with the other objective semantic category, there is little recognition here of how much importance the Russian society assigns to objective descriptions of physical space. This is further supported by the fact that *традиционный* (*traditional*) is found in translated texts as a translation of *traditional* found in the source texts, even though it is not an adjective typically found in Russian tourism discourse.

Example 10.14

ST *You'll find some of Britain's best walking, cycling and horse riding locations and a whole host of other adventure activities in the spectacular National Parks. Add to these 15 National Trails and nearly 50 Areas of Outstanding Natural Beauty (AONBs) and you'll find you're spoilt for choice when it comes to exploring Britain's famous countryside.*

TT *В каждом из них организованы увлекательные маршруты для пешеходных, велосипедных и конных прогулок. А также созданы отличные условия для приключенческого туризма и активного отдыха. И все это – в окружении* **бескрайних долин,** *живописных зелёных холмов, сверкающих озёр и скалистых горных хребтов. Добавьте к этому 15 национальных туристских маршрутов и около 50 зон выдающейся природной красоты – и вопрос, где провести следующий отпуск, отпадёт сам собой.*

BT *Each of them has organized fascinating routes for hiking, cycling and horseback riding. There are also excellent conditions for adventure tourism and active recreation. And all this - surrounded by* ***endless valleys,*** *picturesque green hills, sparkling lakes and rocky mountain ranges. Add to this 15 national tourist routes and about 50 areas of outstanding natural beauty and the question of where to spend your next vacation falls away.*

Source www.visitbritain.com

A similar pattern of lack of adaptation is noted for Beauty adjectives, the frequency of which is almost identical in source and target texts (18.16% vs 18.75%) and higher than that found in Russian non-translated tourism texts (4.49%). This is also manifested by the number of adjective types used in source and target texts (7 vs 6 compared to only 2 found in the non-translations). Although this could also be seen as a consequence of the small corpus component size, since it means that more adjectives with lower frequencies are captured, the frequencies and numbers are so similar between source and target texts that it is very hard to ignore the influence that the former have exerted over the latter. Additionally, Beauty adjectives might occasionally also be added during translation, as seen in Example 10.14, where right after the description of *endless valleys,* green hills are described as *picturesque*. This suggests that there is generally a lack of understanding of what needs to be highlighted to increase the promotional effect of the target texts, when both Size and Beauty adjectives are added. Once again, such additions, although interesting from an analysis point of view, are scarce and do not have a strong impact on the total frequency of Beauty adjectives. Overall, there is an overemphasis on Beauty, especially at the expense of more objective descriptions, which is likely to create an image of the destination that might be considered prejudiced, which can have a negative effect on the promotional potential of such texts.

Finally, no adaptation is noted regarding Fame adjectives either, even though it might be expected. Russian target texts once again follow closely the frequencies found in British English source texts. As a result, Fame adjectives represent a higher frequency in translated Russian texts (8.56%) than non-translated texts (5.07%), closer to what is found in source texts (10.24%). Further evidence of the lack of adaptation is obtained from the use of *легендарный* (*legendary* but also with the meaning of *iconic*) which has not been found in non-translated Russian tourism texts, but *iconic* is found in British English source texts.

Overall, while a high degree of adaptation is required in the case of British English translations into Russian, target texts stay close to their source texts. The resulting texts closely reflect a way of gazing at the world that is ultimately not familiar to the Russian-speaking readership. This might be explained by a perceived status of English (i.e. British), although it is hard to argue with any certainty considering that once again data from only one website are included in the corpus.

10.3.3 Translations into Russian from French

It has been established during the previous phase of analysis that French and Russian tourism texts share many similarities regarding how physical space is described and conceptualised, as evidenced by the lack of statistical significance in the frequencies of the different semantic categories of descriptive adjectives. Therefore, a low degree of adaptation might be expected in this case. While some categories are indeed not adapted, i.e. Size and Fame, adaptation is observed with Quality adjectives, even though it is not required. The only categories where adaptation might be expected are Time and Beauty, but no adaptation is observed in these cases. Overall, there is no clear attempt towards adaptation. Figure 10.9 summarises the quantitative findings presented earlier in Table 10.8.

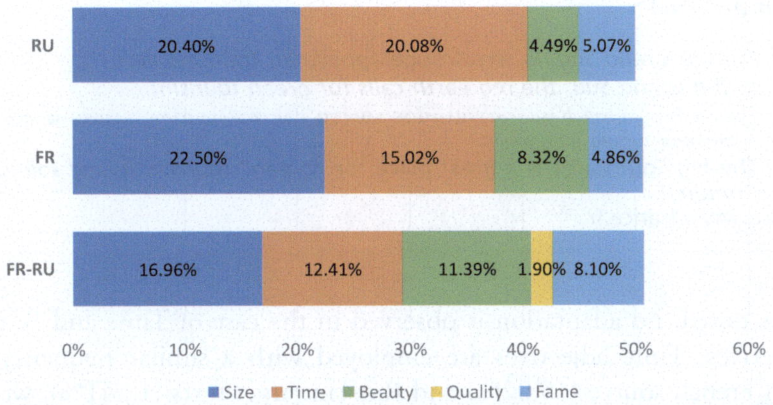

Fig. 10.9 Normalised frequencies of most prominent semantic categories of descriptive adjectives in non-translated Russian, French source texts, and translated Russian from French

Although adaptation might not be expected in the case of Quality adjectives, it is nevertheless noted. No Quality adjectives have been found as being frequently used in Russian or French non-translated tourism texts but they are found in the target texts (1.90%). Example 10.15 illustrates how Quality adjectives might be added, often to translate a particular promotional effect or a wordplay in the source text. This example refers to New Caledonia, an overseas French territory known for the red colour of its ground. The effect created in the source texts by the wordplay between red and green is explicated in the target text with the addition of a Quality adjective. However, it must be noted that the frequency of these adjectives is too low in target texts to conclude with any confidence that they have a meaningful effect on the target reader to the extent that they indeed alter the tourist gaze.

Example 10.15

ST *Dans le Grand Sud, la terre rouge appelle au tourisme vert !*
BT *In the Grand Sud, the red earth calls for green tourism!*
TT *Район Большого Юга* – **отличное место** *для «зеленого» туризма, его красная почва уникальна!*
BT *The Big South area is a* **great place** *for "green" tourism, its red soil is unique!*
Source www.france.fr

As noted, no adaptation is observed in the case of Time and Beauty adjectives. Time adjectives are employed with a similar frequency in both French source (15.02%) and Russian target texts (12.41%), which is lower than what is found in Russian non-translated texts (20.08%). Conversely, the frequency of Beauty adjectives is higher in the target texts (11.39%) compared to Russian non-translated texts (4.49%), reflecting the frequency noted in French source texts (8.32%). There is an under-emphasis on objective descriptions related to chronology and an overemphasis on subjective evaluations of the aesthetics of physical space. This might create the impression in the Russian-speaking reader-tourist that the descriptions are unreliable, as they might not correspond to how they categorise the world around them, which might have an effect on the decision to visit the destination or not.

Overall, as with the other language pairs examined so far, the translations into Russian stay very close to their French source texts, with only minimal differences observed. The resulting target text does not convey an image of the physical that is in-line with how Russian-speaking reader-tourists tend to gaze at the world around them. This is in contrast to what has been observed during the previous case study, where translators from French into Russian in particular seemed to recognise that adaptation is required and make some effort to adjust the target texts accordingly.

10.3.4 Translations into Russian from Greek

Last but not least, although adaptation might be expected for the Greek-Russian language pair due to the statistically significant differences observed during the previous stage of analysis across all semantic categories, no attempt towards adaptation is observed. Figure 10.10 summarises the quantitative findings presented earlier in Table 10.8.

No adaptation is observed in any of the five semantic categories, namely Size, Time, Beauty, Quality, and Fame. Size adjectives in Russian translations (15.83%) are not adapted to match the high frequency with which such adjectives are used in Russian non-translated texts (20.40%), instead they remain close to the frequencies found in Greek source texts (15.29%). This is also manifested by the different Size adjectives employed in the translations, particularly the use of *маленький* (*small*) which is not found in Russian non-translated texts, where its near-synonym *небольшой* (*small*—literally not big) is used instead. This potentially reveals a specific genre preference (i.e. the preference of *маленький* instead of *небольшой* in Russian tourism texts), which does not appear to have been taken into account during translation. Similarly, Time adjectives in Russian translations (11.97%) are also not adapted to match the high frequency with which such adjectives are used in Russian non-translated texts (20.08%) and remain close to the frequencies found in Greek source texts (13.61%). This is also manifested by the presence of the adjective *традиционный* (*traditional*) in

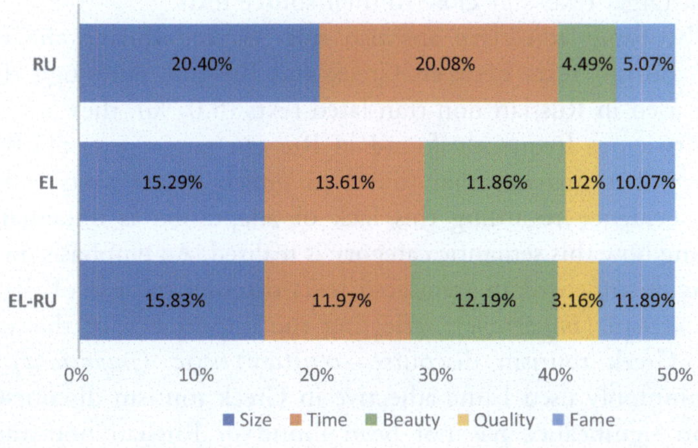

Fig. 10.10 Normalised frequencies of most prominent semantic categories of descriptive adjectives in non-translated Russian, Greek source texts, and translated Russian from Greek

Russian translations, reflecting the use of παραδοσιακός (*traditional*) in the source texts, while this adjective is not typically found in Russian tourism discourse. Overall, there appears to be *less emphasis* on objective descriptions of physical space in the Russian translations from Greek compared to Russian non-translated texts.

This is counterbalanced by a stronger emphasis on subjective descriptions, particularly regarding Beauty adjectives. Although these adjectives represent a relatively low frequency in Russian non-translated texts (4.49%), the frequency found in target texts (12.19%) reflects closely that found in source texts (11.86%). This is also evidenced by the larger variety of Beauty adjectives found in the translations compared to non-translated texts (6 vs 2 types), demonstrating once again influence from Greek source texts, where 5 such adjectives have been found. A similar pattern is observed with the other subjective category, that is, Quality. This category has not been found in Russian non-translated texts, but it is noted in Russian translations from Greek (3.16%) with a similar frequency to that found in Greek source texts (3.12%). The two adjective types found in the target texts, that is *идеальный* (*ideal*) and *хороший* (*good*) are equivalents of the Quality adjectives found in the Greek source texts, that is ιδανικός (*ideal*) and καλός (*good*), further supporting the idea that target texts stay close to their source texts.

Finally, Fame adjectives are also not adapted during the translation of tourism texts between Greek and Russian. Although they are scarcely used in Russian non-translated texts (5.07%), they are approximately twice as frequently found in Russian target texts (11.89%), a frequency that matches that found in Greek source texts (10.07%). Further evidence regarding this lack of adaptation is obtained when examining how this semantic category is realised. An emphasis on significance is documented in translated texts through the use of the adjective *важный* (*important*), reflecting the importance of this concept for the Greek tourism discourse—σημαντικός (*important*) is the most commonly used Fame adjective in Greek tourism discourse. This focus on significance has not been found in Russian non-translated tourism texts. This increases the subjectivity associated with this category, therefore further undermining the objective descriptions that a Russian-speaking audience would prioritise.

Overall, there is, as with all language pairs examined here, a notable lack of adaptation in the case of Russian translations from Greek, even though a high degree of adaptation might be expected. This is in stark contrast to what has been observed during the previous case study (see Chapter 7), where a significant effort towards adaptation has been noted. This results in a target text that shows a different distribution between objective and subjective descriptions of physical space compared to what the Russian-speaking reader-tourist is used to. It is possible that this could create a feeling of insecurity in the target readers, having to rely much more on the subjective evaluations of the author, which might or might not be perceived as accurate. This observation applies to all translations into Russian examined here, which suggests that this lack of adaptation is not particularly affected by the source language.

10.4 Discussion

Despite the many opportunities for adaptation when official tourism websites are being translated into different languages, in practice, such attempts are generally limited, and more importantly they do not appear to be associated with any particular reason, almost giving the impression that whenever adaptation is noted it almost occurs by accident rather than by acknowledgement of its desirability. Nevertheless, even with this general observation in mind, some distinct approaches to translation can be noted, depending on the language pair, similar to what has been noted during the previous case study (Things to do): lack of adaptation, adaptation (to some extent), and an inconsistent approach, mainly where adaptation in the opposite direction is noted.

The first approach, that of lack of adaptation, mainly involves the translations into Russian, particularly from American English and Greek. However, even the translations from British English and French might be considered as belonging to this category, mainly because any attempt towards adaptation (whether expected or not) involves the Quality category, which is not found in Russian non-translated texts, and ultimately represents only a very small frequency in the translated texts, therefore unlikely to have any real impact on the target reader.

It is somewhat surprising that this reluctance to move away from the source text is observed in the case of the Russian translations, even more so when the source text is French or Greek, since prestige cannot be assumed as the reason behind this. Another reason why this reluctance is striking is because Russian tourism texts have been found to clearly stand out from such texts in the other languages (with the possible exception of French). It has been hypothesised that where extreme differences are noted, it is more likely that the need for adaptation will be recognised. This is indeed something that has been observed with some semantic categories for the other language pairs, but not as an overall approach when translating into Russian. It might be that translators have been instructed to follow a close translation or that they themselves have been afraid to move away from the descriptions found in the source texts in case they distort the image of the destination and its promotional effect. If that is the case, which would explain why this lack of adaptation has been noted when differences are most pronounced between languages, it demonstrates a lack of understanding of the importance of adaptation of the tourism message. Let it be noted that translators are not to blame for this state of affairs, even though they are responsible, to a certain extent, for the final text, they might not have received any particular training regarding tourism translation and its particularities, while research on the topic is generally scarce.

Most of the cases where adaptation was seen involved translations into French, as well as most translations into English. However, it must be noted that the degree of adaptation that might be expected and then observed varies significantly. These language pairs simply demonstrate some effort towards adapting at least to some extent how physical space is being described. With the translation into English, Fame appears to be a specific category that is often adapted, while with the translations into French it is typically Size.

It is not clear what the reasons behind these instances of adaptation are. It is possible that, at least for the translations into English, it is important for these texts to appeal to large number of people, considering the number of potential English-speaking reader-tourists. Therefore, translators might feel or be given more freedom to move away from the source text and recreate the message where necessary. It has indeed been observed that where such freedom is present, adaptation tends to be more successful. However, whether or not translators have actually recognised, for instance, the important role that Fame and Size play in English and French tourism discourse, respectively, is hard to tell—it could equally well be a subconscious internalised understanding of what image of the destination would appeal to the target reader. It is worth noting that instances of successful adaptation are relatively few (2 in the case of translations into English and 1 in the case of French translations)—ultimately the target texts are much closer to their source texts than to the respective non-translated texts in the target language.

Finally, sometimes adaptation appears to be inconsistent and without a clear rationale. This has been mainly the case with Greek-English translations. Interestingly, this language pair also exhibited a lack of clear rationale in the previous case study, perhaps the only one to demonstrate a consistent approach, or lack thereof, in both case studies. This is also the only pair where adaptation is noted in the opposite direction to that needed. Such an approach demonstrates a lack of general understanding about the importance of appropriate adaptation during tourism translation and also about how it might look. Sometimes there might be an understanding that changes are required, but not how they need to be implemented. The reason why this has been particularly noted with the translations from Greek might have to do with the fact that these texts are often translated by non-native speakers of English, due to the scarce availability of Greek-English translators. This further confirms the need for more studies such as the present one to help guide translators and provide better understanding of the cross-linguistic differences of tourism discourse.

All in all, for the vast majority of language pairs and semantic categories of adjectives, a lack of adaptation is noted, which confirms the findings from the previous case study that there is a lack of appreciation that adaptation is needed for the tourism message to align with the expectations of a new audience. This does not seem to be affected by whether objective or subjective categories need to be adapted, with evidence of a reluctance to move away from the source texts for both types of adjectives (objective and subjective). This implies that there is also generally little understanding of the importance that objective descriptions can play compared to subjective ones and vice versa, and how these might to some extent be related to the tourism phase, as argued in the previous chapter. This can have significant implications for the resulting target text. On the one hand, readers who are accustomed to gazing at the world through the information value derived from objective descriptions are likely to find a tourism text which employs mainly subjective descriptions unreliable. On the other hand, readers who interpret the world more subjectively and rely on the interpretation provided by the author are likely to find too much emphasis on objective descriptions as unappealing and the resulting texts rather dull.

Similarly, although it has been hypothesised at points that adaptation is more likely to be recognised where differences are more pronounced—something that has also been supported by evidence obtained in the previous case study—there is little evidence from the examination of adjectives to support this. Whether or not societies differ significantly in terms of how they approach physical space appears to have little effect on the decision to adapt or not, which suggests that there is most probably a lack of awareness that such differences might exist. It can even be argued that the target texts aim to present the characteristics of physical space as perceived in the source text because of how closely linked these are to the destination itself (e.g. the use of *ancient* in translations from Greek). However, for a successful promotional effect, it has been claimed that tourism translators need to "meet their target readers' expectations and appeal to culture-bound prejudices" (Smecca 2009, 109). It is indeed questionable whether a different representation of the world can be easily accepted by a different audience, especially if we consider how hard it often is to rebrand other consumer products which might not even have such a close connection to the framing of reality.

Finally, it is interesting, but perhaps not surprising, that when adaptation has been deemed successful it is also when the message has been almost completely rewritten. While an extreme rewriting of the target text might not always be necessary, it goes to show that we need to shift our understanding of tourism texts as general texts that simply need to be accurately translated and that the sole task of the translators is to avoid the notorious translation errors that result in humorous effects. Instead, tourism translation is a specialised type of translation that can also have a significant effect on the promotional potential of the text, and such potential is significantly compromised when the focus is only on accuracy in relation to the source text. This is not only the responsibility of translators, but also of destination promoters, who need to appreciate the important role of translation in this context and ask for translations to fulfil that promotional aim by being more open to changes during that process.

Reference

Smecca, P. 2009. "Tourist Guidebooks and the Image of Sicily in Translation." *Perspectives: Studies in Translatology* 17 (2): 109–19.

11
Final Reflections

It would be easy for someone to argue that this book has revealed nothing new: anyone who has closely examined translations of tourism texts would probably argue that they are rarely extensively adapted and tend to convey the same message as their source texts. If we limit the contribution of this book to this statement, then we have missed its significant potential to challenge current thinking and practices. The main contribution of this book is that, although it does not aim to answer them directly, it provides some insights into two important questions: why tourism texts should be adapted in many cases and why data examined in the two case studies indicate that they are not.

Throughout the book, I have shown the importance of adaptation in the language of tourism, drawing strong parallels between advertising and tourism promotion. Even though such parallels might appear obvious to some readers, they are not always recognised by tourism translators. There is significant scope in expanding this area of research to highlight even further the promotional nature and potential of tourism texts. More importantly though, the two case studies presented in this book also explore the importance of adaptation in tourism promotion by

S. Malamatidou, *Translating Tourism*, Palgrave Studies in Translating and Interpreting, https://doi.org/10.1007/978-3-031-49349-2_11

relying on concrete concepts of how tourism is realised and what considerations guide and formulate tourist behaviour and destination image. Unlike many previous studies, the case studies presented in this book do not provide a list of linguistic features that happen to differ cross-linguistically. After all, such a list would be of little value, as it is almost impossible to remember all the differences, without understanding why they exist. Instead, the aim is to link specific linguistic choices to the way people (i.e. destination promoters) use these to communicate a specific message around tourism, which is related to how a particular society experiences tourism (i.e. tourism phase) and what elements of it are prioritised (i.e. tourism gaze), which are bound to differ cross-culturally. This is one of the first studies to approach tourism discourse in such a way and it is hoped that it will pave the way for similar studies in the future, inspiring a long-overdue engagement with interdisciplinary research in this area.

Results from the two case studies in this book reveal that, although some similarities are observed, the different languages examined here employ specific linguistic means (i.e. verbs and adjectives) differently. These reveal that the societies associated with these languages perceive the physical world around them differently; they also experience tourism differently. At the same time, the data reveal that we can indeed talk about a universal language of English tourism, since differences between American and British English tourism texts have been found to be minimal in those aspects analysed here. However, where differences are noted, they do not seem to be taken into account during translation across the different language pairs examined here. Even when adaptation is observed, it is not systematic, which suggests that it might also have occurred by chance. The general tendency is for the target texts to stay close to their source texts, and for semantic equivalence to be given priority over pragmatic effect and cross-cultural differences in the way different societies experience and view the world around them. This can have significant implications for the promotional potential of the tourism message, since the resulting target texts do not appear to (re)create an image of the destination, and tourism more generally that is in line with the expectations of the reader-tourists. This can potentially result in the tourism message being perceived as weak or unappealing. Translation has been found to have missed an opportunity to become a field of active destination identity reconstruction.

These findings provide further evidence regarding the importance of adaptation in the translation of tourism texts. Put simply, tourism texts need to be adapted during translation so that they can attract more potential visitors and increase their promotional potential. More specifically, they need to be adapted so that they are able to reflect the version of the world that is compatible with the target society or culture in which they will circulate. While we know very little about the effects that a potential incompatibility due to lack of adaptation, as it is currently manifested in translated texts, might have on the target audience, it can be hypothesised that, apart from the obvious compromise of the promotional potential of tourism texts, it can also affect our understanding of the world. When contradicting worldviews circulate, this can significantly compromise the normal processes of tourism development. For instance, it can potentially have some influence (together with other factors) on a society to reach the next stage of tourism or alternatively regress into the previous one. Similarly, it can result in a state of psychological confusion, where receivers of tourism texts are unsure about what really constitutes tourism. This is not unlike the sense of information overload that people can experience nowadays. Similarly, if we examine tourist gaze, multiple 'gazes' can also result in a state of confusion or even, in extreme cases, in a shift in tourist gaze which might or might not be desirable.

Related to desirability, the discussion around the need to adapt tourism texts has also confirmed the power of language in destination promotion. While it is easy to dismiss it as simply promotional discourse focusing only on economic gain, and therefore having little intellectual value, in reality its potential in shaping behaviours and attitudes is significant. Understanding which linguistic features might encode which aspects of the tourism experience is vital in manipulating these in order to shift perceptions and stereotypes, if that is something that needs to be achieved. This is relevant to the initial production of tourism texts, before translation is involved, but can of course extend to translation. Although it has been argued that cliches and stereotypes are hard to shift in tourism discourse (Dann 2001), it is still possible to talk about a 'controlled promotion' (Malamatidou 2018) where at least some amount of reflection has gone into how a message has been shaped. This becomes even more relevant if we consider the interconnection between tourism and what is happening in the rest of the society or even the world, as already discussed in the Introduction to this book. For example, the discourse of tourism changed in some ways during the COVID-19 pandemic to focus much more on safety as a priority. While this is a rather obvious example, it shows the potential that tourism discourse has. A question therefore naturally arises around how we could harvest such potential to create different attitudes around pressing issues like over-tourism or a shift towards greener tourism, which is something that tourism studies will hopefully be able to answer in the future.

The book also provides insight into why tourism texts might not be adapted during translation. Or to be more specific, what the reasons are for an evident reluctance to adapt tourism translation. We can speculate that there are many reasons, some of which might constitute a conscious, yet still badly informed, decision, while others might be subconscious, but also relying on a lack of information. There is one key element here: knowledge about what needs to be adapted and, even more importantly, why. As already discussed in previous chapters, it is not easy to pinpoint who exactly lacks this knowledge, whether it is translators or destination promoters as commissioners of the translations. In other words, it is unclear whether translators purposefully produce texts that stay close to source text meaning, or whether commissioners are reluctant to accept a project where significant changes regarding meaning have been made. More research is needed in this area, but this book provides, for the first time, some key information around the importance of paying more attention to how tourist behaviour is realised through the use of specific verbs and how this is linked to the tourism phase a society is experiencing, as well as to how destinations are framed through the use of specific adjectives and how this is linked to the concept of tourist gaze. This information can prove valuable to both translators and destination promoters, highlighting how truly disciplinary tourism translation is.

Interdisciplinarity in this case is not restricted to educating both translators and commissioners about the particularities of tourism translations, although this is indeed a very important element of it. It also refers to understanding how concepts of tourism studies, such as the tourism phase and tourist gaze examined here, can significantly enrich our understanding of cross-cultural communication. Tourism, which by definition allows different cultures to come together and is all about describing the world around us (or rather a version of it), is the perfect springboard for gaining insight into how humans experience and perceive the world and how this might differ across different cultures and linguistic groups. Studying tourism discourse and its translation not only enriches our understanding of translation and cross-cultural communication, it also provides a window into how we humans try to make sense of the world around us, how we categorise its elements, and how we see ourselves within it. Tourism translation is, I hope this book has shown, much more than amusing errors—it allows us to challenge some of the assumptions we have about the world and the stereotypes we assume. However, this can only be achieved if we contextualise meaning and gain an understanding of the mechanisms that drive tourism activity.

We are at the very beginning of this enriching journey, as we need more synergies between tourism and translation studies to appreciate the full potential of tourism discourse. It is hard to envisage where this journey might take us, but one much desired outcome would be to transform translators into true cultural consultants when it comes to tourism, focusing on both words and worlds. This will of course require research into tourism translation, training translators, and educating destination promoters, which needs time. But I believe the first steps have already been made and that this book makes a significant contribution towards that aim. I hope that this book has not only answered questions but has also created new ones, paving the way for future investigations.

References

Dann, G. 2001. "The Self-Admitted Use of Cliché in the Language of Tourism." *Culture & Communication* 3: 1–14.

Malamatidou, S. 2018. "'A Pretty Village Is a Welcome Sight': A Contrastive Study of the Promotion of Physical Space in Official Tourism Websites." *Translation Spaces* 7 (2): 304–33.

Appendix I

Results of statistical tests for the distribution of most prominent verbs in non-translated tourism texts

	EN_US vs EN_UK	EN_US vs FR	EN_US vs EL	EN_US vs RU	EN_UK vs FR	EN_UK vs EL	EN_UK vs RU	FR vs RU	EL vs FR	EL vs RU
Motion										
LL	10.75	4.26	2.92	54.34	27.94	20.69	101.07	30.82	0.03	29.27
BIC	1.78	−4.21	−5.44	46.17	19.07	11.89	92.39	22.88	−8.14	21.52
Discovery										
LL	0.06	6.77	35.24	86.76	9.82	45.46	91.99	161.23	11.01	161.23
BIC	−8.91	−1.7	26.88	78.78	0.96	36.67	83.3	153.49	2.84	153.49
Perception										
LL	7.11	0.09	7.44	51.9	4.27	0.66	34.9	44.44	5.31	161.23
BIC	−1.85	−8.38	−0.92	43.72	−4.6	−4.88	26.22	36.49	−2.86	153.49
Emotion										
LL	1.75	5.39	4.4	N/A	14.48	11.48	N/A	N/A	0	N/A
BIC	−7.22	−3.08	−3.96		5.61	2.69			−8.17	

S. Malamatidou, *Translating Tourism*, Palgrave Studies in Translating and Interpreting, https://doi.org/10.1007/978-3-031-49349-2

Results of statistical tests for the distribution of most prominent verbs in English translated tourism texts

	FR-EN T vs EN_ US NT	FR-EN T vs EN_ UK NT	FR-EN T vs FR NT	EL-EN T vs EN_ US NT	EL-EN T vs EN_ UK NT	EL-EN T vs EL NT	RU-EN T vs EN_US NT	RU-EN T vs EN_ UK NT	RU-EN T vs RU NT
Motion									
LL	7.68	39	0.34	55.55	123.6	24.15	67.76	119.42	0.6
BIC	−0.84	30.09	−8.03	47.16	114.78	16.1	59.6	110.74	−6.76
Discovery									
LL	0.4	0.23	9.5	0.42	0.85	22.48	22.5	23.3	15.04
BIC	−8.13	−8.68	1.13	−7.97	−7.97	14.42	14.34	14.63	7.67
Perception									
LL	7.71	31.18	8.21	26.25	59.83	48.37	12.37	4.17	7.69
BIC	−0.82	22.27	−0.16	17.86	51.02	40.31	4.21	−4.51	0.33
Emotion									
LL	12.53	27.82	1.1	7.38	17.1	0.23	6.22	3.39	N/A
BIC	4.01	18.91	−7.27	−1.01	8.28	−7.83	−1.95	−5.29	

Results of statistical tests for the distribution of most prominent verbs in French translated tourism texts

	EN_ US-FR T vs FR NT	EN_ US-FR T vs EN_ US NT	EN_ UK-FR T vs FR NT	EN_ UK-FR T vs EN_ UK NT	EL-FR T vs FR NT	EL-FR T vs EL NT
Motion						
LL	0.01	5.39	17.53	139.97	4.47	4.61
BIC	−8.65	−3.38	8.88	130.88	−3.49	−3.16
Discovery						
LL	16.4	1.97	26.32	7.43	17.31	1.65
BIC	7.75	−6.81	17.67	−1.66	9.34	−6.12
Perception						
LL	2.15	1.59	0.34	3.2	0.45	1.43
BIC	−6.51	−7.19	−8.31	−5.88	−7.52	−6.34
Emotion						
LL	7.76	0.12	21.11	1.54	2.62	2.32
BIC	−0.9	−8.65	12.45	−7.54	−5.35	−5.45

Results of statistical tests for the distribution of most prominent verbs in Russian translated tourism texts

	EN_ US RU T vs RU NT	EN_ US RU T vs EN_ US NT	EN_ UK RU T vs RU NT	EN_ UK RU T vs EN_ UK NT	FR RU T vs RU NT	FR RU T vs FR NT	EL RU T vs RU NT	EL RU T vs EL NT
Motion								
LL	62.45	3.49	3.31	71.33	2.31	13.47	1.55	17.28
BIC	54.8	−4.81	−4.21	62.61	−5.07	5.52	−5.9	9.5
Discovery								
LL	27.95	19.78	1.18	87.72	7.41	59.63	28.68	56.54
BIC	20.31	11.48	−6.34	79	0.03	51.69	21.23	48.75
Perception								
LL	31.75	0.8	11.98	2.72	3.76	18.11	0.14	17.84
BIC	24.1	−7.5	4.46	−6	−3.62	10.17	−7.31	10.05
Emotion								
LL	N/A	3.33	N/A	8.18	N/A	19.7	N/A	19.06
BIC		−4.97		−0.54		11.75		11.27

Appendix II

Results of statistical tests for the distribution of most prominent descriptive adjectives in non-translated tourism texts

	EN_ US vs EN_ UK	EN_ US vs FR	EN_ US vs EL	EN_ US vs RU	EN_ UK vs FR	EN_ UK vs EL	EN_ UK vs RU	FR vs EL	FR vs RU	EL vs RU
Size										
LL	8.29	79.09	22.28	59.15	156.06	82.83	140.05	26.58	0.58	18.85
BIC	−0.06	70.96	13.68	51	147.66	74.06	131.63	17.95	−7.63	10.2
Time										
LL	58.72	101.84	103.18	183.81	11.86	7.14	56.9	1.69	13.48	32.31
BIC	50.37	93.72	94.58	175.65	3.46	−1.63	48.48	−6.94	5.27	23.66
Beauty										
LL	9.37	8.09	0.03	8.49	38.4	258.29	75.58	9.79	7.56	35.69
BIC	1.27	0.34	−8.21	0.65	30.31	249.82	67.55	1.56	−0.09	27.51
Quality										
LL	4.5	N/A	92.07	N/A	N/A	N/A	N/A	N/A	N/A	N/A
BIC	−3.59		83.83							
Fame										
LL	7.98	69.28	10.18	66.73	40.9	0.04	38.43	35.64	3.88	11.46
BIC	33.85	4.09	−7.63		−0.82	45.96		2.77		

S. Malamatidou, *Translating Tourism*, Palgrave Studies in Translating and Interpreting, https://doi.org/10.1007/978-3-031-49349-2

Results of statistical tests for the distribution of most prominent descriptive adjectives in English translated tourism texts

	FR-EN T vs EN_ UK NT	FR-EN T vs EN_US NT	FR-EN T vs FR NT	EL-EN T vs EN_ UK NT	EL-EN T vs EN_US NT	EL-EN T vs EL NT	RU-EN T vs EN_UK NT	RU-EN T vs EN_US NT	RU-EN T vs RU NT
Size									
LL	162.29	73.64	0.06	156.83	59.16	16.47	83.39	32.68	1.42
BIC	153.92	65.53	−8.1	148.11	50.63	7.57	75.12	24.71	−6.64
Time									
LL	38.54	153.08	6.53	30.99	152.95	10.68	83.34	213.6	4.84
BIC	30.16	144.98	−1.63	22.26	144.41	1.78	75.07	205.64	−3.23
Beauty									
LL	17.15	12.33	18.66	88.05	0.13	10.89	158.58	28.22	0.5
BIC	8.77	4.23	10.5	79.32	−8.4	1.99	150.31	20.26	−7.56
Quality									
LL	10.64	10.6	N/A	157.97	128.88	9.2	N/A	N/A	N/A
BIC	2.26	2.49		149.25	120.34	0.3			
Fame									
LL	0.25	9.08	29.27	0.32	6.17	0.72	0.17	7.52	23.87
BIC	−8.12	0.98	21.11	−8.4	−2.36	−8.18	−8.11	−0.45	15.81

Results of statistical tests for the distribution of most prominent descriptive adjectives in French translated tourism texts

	EN_US-FR T vs FR NT	EN_US-FR T vs EN_ US NT	EN_UK-FR T vs FR NT	EN_UK-FR T vs EN_ UK NT	EL-FR T vs FR NT	EL-FR T vs EL NT
Size						
LL	20.02	19.05	65.56	21.38	0.22	29.61
BIC	11.89	10.99	57.32	12.94	−8	20.96
Time						
LL	66.27	4.01	36.42	9.44	0.24	3.65
BIC	58.14	−4.06	28.18	1	−7.97	−5
Beauty						
LL	1.16	0.15	69.38	0	7.63	0.55
BIC	−6.97	−7.92	61.14	−8.44	−0.58	−8.1
Quality						
LL	N/A	16.84	N/A	11.97	N/A	3.33
BIC		8.77		3.53		−5.32
Fame						
LL	79.69	0.35	151.62	50.71	75.76	11.89
BIC	71.56	−7.72	143.38	42.27	67.54	3.24

Results of statistical tests for the distribution of most prominent descriptive adjectives in Russian translated tourism texts

	EN_ US-RU T vs RU	EN_ US-RU T vs EN_ US	EN_ UK RU T vs RU	EN_ UK RU T vs EN_ UK	FR-RU T vs RU	FR-RU T vs FR	EL-RU T vs RU	EL-RU T vs EL
Size								
LL	74	6.12	46.16	5.82	3.49	8.4	11.99	0.27
BIC	66.09	−1.68	38.18	−2.39	−4.4	0.55	3.65	−8.45
Time								
LL	133.86	0.76	50.01	3.91	19.79	2.7	43.24	2.99
BIC	125.95	−7.04	42.03	−4.3	11.9	−5.15	34.91	−5.73
Beauty								
LL	17.82	0.01	134.93	0.14	36.87	5.38	75.25	0.13
BIC	9.9	−7.78	126.94	−8.07	28.98	−2.47	66.91	−8.58
Quality								
LL	N/A	1.66	N/A	82.34	N/A	N/A	N/A	0.01
BIC		−6.13		74.13				−8.71
Fame								
LL	30.29	1.69	12.4	2.22	8.08	9.31	57.48	4.3
BIC	22.38	−6.11	4.41	−5.99	0.19	1.45	49.14	−4.41

Index

S. Malamatidou, *Translating Tourism*, Palgrave Studies in Translating and Interpreting, https://doi.org/10.1007/978-3-031-49349-2

S

T

SPRINGER NATURE

GPSR Compliance

The European Union's (EU) General Product Safety Regulation (GPSR) is a set of rules that requires consumer products to be safe and our obligations to ensure this.

If you have any concerns about our products, you can contact us on ProductSafety@springernature.com

In case Publisher is established outside the EU, the EU authorized representative is:

Springer Nature Customer Service Center GmbH
Europaplatz 3
69115 Heidelberg, Germany

The manufacturer's authorised representative in the EU is Springer Nature Customer Service Centre GmbH, Europaplatz 3, 69115 Heidelberg, Germany. If you have any concerns regarding our products, please contact ProductSafety@springernature.com

Printed and bound by CPI Group (UK) Ltd, Croydon, CR0 4YY
01/10/2026
02225898-0008